SHOW ME THE NUMBERS

SHOW ME THE NUMBERS

Designing Tables and Graphs to Enlighten

SECOND EDITION

STEPHEN FEW

Analytics Press

BURLINGAME, CALIFORNIA

Analytics Press

PO Box 1545
Burlingame, CA 94011
SAN 253-5602
www.analyticspress.com
Email: info@analyticspress.com

PUBLISHER: Jonathan G. Koomey

COPY EDITOR: Nan Wishner
COMPOSITION: Bryan Pierce
COVER DESIGN: Stephen Few
PHOTOGRAPHY: John Fernez
PRINTER AND BINDER: C&C Offset Printing Company

ISBN-10: 0-9706019-7-2
ISBN-13: 978-0-9706019-7-1

This book was printed on acid-free paper in China.

10 9 8 7 6

If we've done well in life, we have fellow travelers to thank. No one succeeds alone. In my life's story, many of the heroes are teachers. Several extraordinary educators from elementary school through graduate school invited me into a world of never-ending wonder and gave me the confidence to make it my own. More than the content of their lessons, what influenced me most were the ways they found to connect with me as a person who was worthy of their time. Their remarkable kindness, commitment to education, and practiced skill helped me become the person, and the teacher, that I am today. If I've done and can continue to do for others what these teachers have done for me, my life has meaning.

Oh, the thirst to know
how many!
The hunger
to know
how many
stars in the sky!
We spent
our childhood counting
stones and plants, fingers and
toes, grains of sand, and teeth,
our youth we passed counting
petals and comets' tails.
We counted
colors, years,
lives, and kisses;
in the country,
oxen; by the sea,
the waves. Ships
became proliferating ciphers.
Numbers multiplied.
The cities
were thousands, millions,
wheat hundreds
of units that held
within them smaller numbers,

smaller than a single grain.
Time became a number.
Light was numbered
and no matter how it raced with sound
its velocity was 37.
Numbers surrounded us.
When we closed the door
at night, exhausted,
an 800 slipped
beneath the door
and crept with us into bed,
and in our dreams
4000s and 77s
pounded at our foreheads
with hammers and tongs.
5s
added to 5s
until they sank into the sea or madness,
until the sun greeted us with its zero
and we went running
to the office,
to the workshop,
to the factory,
to begin again the infinite
1 of each new day.

Excerpt from the poem
"Ode to Numbers"
by Pablo Neruda,
Selected Odes of Pablo Neruda,
translated by Margaret Sayers Peden
(1995). University of California Press.

CONTENTS

14. **THE INTERPLAY OF STANDARDS AND INNOVATION**
When you design tables and graphs, you face many choices. Of the available alternatives, some are bad, some are good, some are best, and others are simply a matter of preference among equally good choices. By developing and following standards for the visual display of quantitative information, you can eliminate all but good choices once and for all. This dramatically reduces the time it takes to produce tables and graphs as well as the time required by your readers to make good use of them. Doing this will free up time to put your creativity to use where it's most needed.

PREFACE TO SECOND EDITION

In 1914, ninety years before the first edition of this book was published, Willard C. Brinton wrote what was perhaps the first book about graphical data presentation, entitled *Graphic Methods for Presenting Facts*. The relatively few books on the topic that have been published since Brinton's have mostly appeared during the last 20 years. If you read Brinton's trailblazing book, you will be surprised by how little has changed since he wrote it. The problems that he tried to solve were not so different from those that we face today. Brinton began his book with the following words:

> After a person has collected data and studied a proposition with great care so that his own mind is made up as to the best solution for the problem, he is apt to feel that his work is about completed. Usually, however, when his own mind is made up, his task is only half done. The larger and more difficult part of the work is to convince the minds of others that the proposed solution is the best one—that all the recommendations are really necessary. Time after time it happens that some ignorant or presumptuous member of a committee or a board of directors will upset the carefully-thought-out plan of a man who knows the facts readily enough to overcome the opposition. It is often with impotent exasperation that a person having the knowledge sees some fallacious conclusion accepted, or some wrong policy adopted, just because known facts cannot be marshaled and presented in such manner as to be effective.
>
> Millions of dollars yearly are spent in the collection of data, with the fond expectation that the data will automatically cause the correction of the conditions studied. Though accurate data and real facts are valuable, when it comes to getting results the manner of presentation is ordinarily more important than the facts themselves. The foundation of an edifice is of vast importance. Still, it is not the foundation but the structure built upon the foundation which gives the result for which the whole work was planned. As the cathedral is to its foundation so is an effective presentation of facts to the data.[1]

1. Willard C. Brinton (1914) *Graphic Methods For Presenting Facts*. The Engineering Magazine Company, pages 1 and 2.

Accumulating information in and of itself is not useful. Information can't possibly serve a purpose until we first identify what's meaningful and then manage to make sense of it. Even once we understand the information, it remains inert until we actually do something with it. The true promise of the information age isn't tons of data but decisions and actions that are better because they're based on an understanding of what's really going on in the world. Any knowledge that you gain that could be used to make better decisions will amount to nothing if you can't communicate it to others in a way that makes sense to them. The ability to find what's useful in the mounds of data that surround us, to make sense of it, and to then present it clearly and accurately, forms the foundation on which the information age will finally fulfill its promise. Unless we give information a clear voice, its important stories will remain unheard, and ignorance will prevail.

As I write these words, more than 10 years into the 21st century, I still feel compelled to make the same essential case that Brinton made long ago when he wrote:

> *If an editor should print bad English he would lose his position. Many editors are using and printing bad methods of graphic presentation, but they hold their jobs just the same. The trouble at present is that there are no standards by which graphic presentations can be prepared in accordance with definite rules so that their interpretation by the reader may be both rapid and accurate. It is certain that there will evolve for methods of graphic presentation a few useful and definite rules which will correspond with the rules of grammar for the spoken and written language. The rules of grammar for the English language are numerous as well as complex, and there are about as many exceptions as there are rules. Yet we all try to follow the rules in spite of their intricacies. The principles for a grammar of graphic presentation are so simple that a remarkably small number of rules would be sufficient to give a universal language.*[2]

2. *ibid.*, page 3.

Even though no precedent for codifying the rules of graphic presentation existed in his day, Brinton made a bold and brilliant attempt to begin this work. Nearly a century has now passed, and in that time I and others have continued the work, resulting in the simple set of rules that Brinton hoped for, but these rules are known by relatively few who need them. In Brinton's time, the computer was still many years in the future. Graphics were produced by hand, usually by professional draftsmen. It is sad that since the advent of the computer—especially personal computing, which gave everyone the means to produce graphs and to do so efficiently—the quality of graphical communication has actually diminished. Having the means to create graphs with a computer doesn't guarantee that we'll do it effectively any more than having word processing software makes us great writers. It seems that we've lost sight of this distinction and assume that if we know how to use software that was designed to produce graphs, we have all that we need. Relying on software to do this for us results in failure. Software can do little to help us communicate graphically if we don't already possess the basic skills to do it ourselves.

You might be wondering why I'm writing a new edition of this book only eight years after the first edition was published. Has that much changed? Although relatively few changes have taken place in data visualization, I've learned a great deal more than I knew eight years ago. Most of the changes that I've made to this new edition are the results of my own professional growth.

Show Me the Numbers was the first of three books that I've now written. When I wrote it originally I had only recently begun focusing on data visualization despite having worked for nearly 20 years in information technology, mostly in *business intelligence*. I devoured the work of Edward Tufte and others, compiled the best of it, ran it through the filter of my own experience, then organized and expressed it to address the practical needs of people like you. Since writing the first edition of this book, I have taught data visualization courses internationally to thousands of people, written scores of articles, white papers, and two more books, and have worked with and advised many diverse organizations. As a result, my expertise has matured. This second edition of *Show Me the Numbers* is the result of this maturity.

In addition to my professional development, a few things have been happening in the world that affect graphical communication—some positive, which I describe in this edition, and some negative, which I warn against. On the positive side, two people in particular have shown that important stories involving numbers can be told in compelling ways. Even though Al Gore did not invent data visualization (or the Internet), the compelling nature of the graphical displays in his film *An Inconvenient Truth* began to change the tide of opinion about global warming. Those graphical displays moved people thanks to expert assistance from Nancy Duarte and Duarte Design. Another person who has used graphics to capture the imaginations of many in recent years is Hans Rosling of GapMinder.org. When this Swedish professor took the stage at the 2006 Technology, Entertainment, and Design (TED) Conference and told a story about the relationship between fertility and life expectancy throughout the world from 1962 to the present using an animated bubble chart, a new era of quantitative storytelling began.

On the negative side, the availability of bad graphs has increased since the first edition of this book. This expansion of bad graphical presentation has been made possible by the Web and has been fueled by uninformed so-called experts and self-serving software vendors. During the past few years, the number of people who claim expertise in data visualization has increased dramatically, but unfortunately many of them do not exhibit best practices. This is especially true of many graphic artists whose data-based visualizations are greeted with fanfare even though they don't actually inform, or do so poorly. Flashy visual displays are engaging, but unless they invite people to think about data in meaningful ways that lead to understanding, they fail in their purpose.

Nothing has been more disappointing to me personally than the lack of improvement in the charting capabilities of the product that is used more than any other to produce graphs: Microsoft Excel. The 2007 and 2010 releases of Excel have added superficial sizzle to the product's graphs without addressing any of the fundamental charting problems that have existed for years. Excel still encourages people to produce bad graphs—in some respects more than in the past because it now offers even more dysfunctional choices. Nevertheless, because practically everyone in the world who produces graphs has a copy of Excel, I've made a point of featuring graphs in this book that can be created effectively with Excel if you know what you're doing. Fortunately, the necessary skills are quite simple and easy to learn.

The purpose of this book is to help people present quantitative information in the most informative way possible, using simple skills and tools that are readily available. I invite you to enjoy the journey of learning these essential skills.

1   INTRODUCTION

The use of tables and graphs to communicate quantitative information is common practice in organizations today, yet few of us have learned the design practices that make them effective. This introductory chapter prepares the way for a journey of discovery that will enable you to become an exception to this unfortunate norm.

"Show me the numbers" is an expression that I've heard from time to time, especially in the workplace. Numbers are central to our understanding of performance. They enable us to make informed decisions. The way we determine success or failure is almost always based primarily on numbers. We derive great value from the stories that numbers tell, yet we rarely consider the significance of how we present them.

Contrary to popular wisdom, numbers cannot always speak for themselves. Inattention to the display of quantitative data results in large but hidden costs to most organizations. Time is wasted struggling to understand the meaning and significance of numbers—time that could be better spent doing something about them. What a shame, especially when this could be easily remedied. To provide a practical solution to this pervasive problem is the goal of this book.

Leveraging quantitative information—numbers—takes us out of the realm of assumption, feeling, guesswork, gut instinct, intuition, and bias, and into the realm of reliable fact based on measurable evidence. Too many decisions are based on perceptions that are fallible. You may wake up in the morning, step outside, feel the sunshine on your skin, and know deep down in your bones that it's warmer today than it was yesterday, only to glance at the thermometer and discover that it is in fact five degrees cooler. More to the point of this book, your gut may tell you that your organization is now doing better than ever, but a careful check of the numbers could reveal that performance has actually degraded during the past 12 months.

I respect the value of well-honed intuition. I personally trust a strong intuitive sense of what's going on, of what will work and what won't, that is rarely wrong when applied to my areas of expertise. Rarely wrong, but wrong often enough to keep me humble. By trusting my gut—overly confident that the extra step of actually checking the numbers would be wasted—I have at times been dragged kicking and screaming to the embarrassing admission that I was wrong. Numbers have an important story to tell, and it is up to us to help them tell it. Tables and graphs are usually the best means to communicate quantitative information. They are so common many of us assume that knowledge of their effective use is common as well. I assure you, it is not. Evidence of this fact in the form of countless poorly designed tables and graphs is visible all around us. According to Karen A. Schriver, an expert in document design, "Poor documents are so commonplace that deciphering bad writing and bad visual design have become part of the coping skills needed to navigate in the so-called information age."[1]

1. Karen A. Schriver (1997) *Dynamics in Document Design*. John Wiley & Sons, Inc., page xxiii.

The reason for this sad state of affairs is simple: few of us have ever been trained to design tables and graphs effectively. Some of us have struggled to do this work well but haven't found useful resources to assist us. Others of us haven't struggled at all, simply because we haven't seen enough examples of good design to recognize the inadequacy of our own efforts.

Before saying anything more, let me illustrate what I mean. Take a moment to examine the graph below.

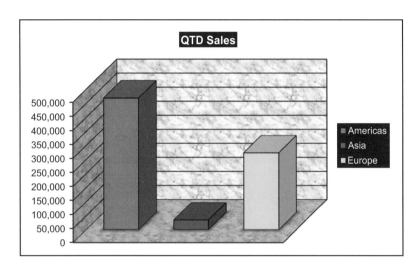

FIGURE 1.1 This is a typical example of a poorly designed business graph. Notice the attempt at artistic flair in the use of color, 3D, and shading of the vertical bars.

This graph has visual impact—it's dramatic, it's colorful, it jumps off the page and demands attention—but to what end? What is the message?

Let's evaluate its effectiveness in light of the objective. Assume that its purpose is to inform a corporation's executives every Monday morning about the current state of quarterly sales, split into three geographical regions: Americas, Europe, and Asia. Given this intention, look at it again. What message could an executive get from this graph? Put yourself in the executive's position. Take a minute or two to examine the graph and interpret its message in this light.

. . . .

What did you get? Probably no more than the following:

- Sales for the Americas are better than sales for Europe, which in turn are better than sales for Asia. This much is clear.
- Sales for Asia don't amount to very much compared to the other regions.
- Sales for the Americas are more than 400,000, or thereabouts. Sales for Europe are somewhere around 200,000. Sales for Asia appear to be around 50,000, perhaps a little less.

That's not much information, and you certainly had to work for it, didn't you? Here are the actual values that were used to create this graph:

- Americas = $469,384
- Europe = $273,854
- Asia = $34,847

As an executive, you may not need to know precise sales amounts, but you would probably want greater accuracy than you could discern from this graph, and you would certainly want to get it faster and with less effort.

Several pieces of critical information aren't supplied by this graph:

- Given the fact that these sales are international, what is the unit of measure? Is it U.S. dollars, Euros, etc.?
- What is the date of the "quarter-to-date sales?" If you filed this report away for future reference and pulled it out again a year from now, you wouldn't know what day, what quarter, or even what year it represents.
- How do these sales figures compare to your plan for the quarter?
- How do these sales figures compare to how you did at this time last quarter or this same quarter last year?
- Are sales getting better or worse?

This graph lacks important contextual information and critical points of comparison. As a report of quarter-to-date sales across your major geographical regions, it doesn't show you very much. It uses a great deal of ink to say very little.

Given the intended message and the information an executive might find useful, the following table tells the story better:

2011 Q1-to-Date Regional Sales
As of March 15, 2011

	Current			Qtr End	
	Sales (U.S. $)	Percent of Total Sales	Percent of Qtr Plan	Projected Sales (U.S. $)	Projected Percent of Qtr Plan
Americas	469,384	60%	85%	586,730	107%
Europe	273,854	35%	91%	353,272	118%
Asia	34,847	5%	50%	43,210	62%
	$778,085	100%	85%	$983,212	107%

Note: To date, 82% of the quarter has elapsed.

FIGURE 1.2 This table contains all of the information that was contained in the graph in *Figure 1.1*, plus much more, in an easy-to-grasp format.

This table is simple and easy to read, yet it contains a great deal more information than the previous, colorful graph. No ink is wasted. As an executive, you might actually be able to use this report to make important decisions. It communicates.

Here's some information that I found on the Web that was featured by a PBS television program called NOW:

Favorable or Unfavorable View of the U.S.	
Brazil: % with somewhat or very favorable opinion of the U.S.:	52%
Brazil: % with somewhat or very unfavorable opinion of the U.S.:	32%
Mexico: % with somewhat or very favorable opinion of the U.S.:	64%
Mexico: % with somewhat or very unfavorable opinion of the U.S.:	25%
Britain: % with somewhat or very favorable opinion of the U.S.:	75%
Britain: % with somewhat or very unfavorable opinion of the U.S.:	16%
Germany: % with somewhat or very favorable opinion of the U.S.:	61%
Germany: % with somewhat or very unfavorable opinion of the U.S.:	35%
Russia: % with somewhat or very favorable opinion of the U.S.:	61%
Russia: % with somewhat or very unfavorable opinion of the U.S.:	33%
Poland: % with somewhat or very favorable opinion of the U.S.:	79 %
Poland: % with somewhat or very unfavorable opinion of the U.S.:	11%
South Africa: % with somewhat or very favorable opinion of the U.S.:	65%
South Africa: % with somewhat or very unfavorable opinion of the U.S.:	28%
Kenya: % with somewhat or very favorable opinion of the U.S.:	80%
Kenya: % with somewhat or very unfavorable opinion of the U.S.:	15%
India: % with somewhat or very favorable opinion of the U.S.:	54%
India: % with somewhat or very unfavorable opinion of the U.S.:	27%
Japan: % with somewhat or very favorable opinion of the U.S.:	72%
Japan: % with somewhat or very unfavorable opinion of the U.S.:	26%
South Korea: % with somewhat or very favorable opinion of the U.S.:	53%
South Korea: % with somewhat or very unfavorable opinion of the U.S.:	44%
Egypt: % with somewhat or very favorable opinion of the U.S.:	6%
Egypt: % with somewhat or very unfavorable opinion of the U.S.:	69%
Pakistan: % with somewhat or very favorable opinion of the U.S.:	10%
Pakistan: % with somewhat or very unfavorable opinion of the U.S.:	69%
Turkey: % with somewhat or very favorable opinion of the U.S.:	30%
Turkey: % with somewhat or very unfavorable opinion of the U.S.:	55%
Jordan: % with somewhat or very favorable opinion of the U.S.:	25%
Jordan: % with somewhat or very unfavorable opinion of the U.S.:	75%

FIGURE 1.3 This table reports information that was collected by the Pew Research Center, as presented by the PBS television program NOW.

When I first looked at this information not long after the horrific events of September 11, 2001, I was frustrated. I wanted to understand how these countries felt about America, but I couldn't get a clear picture from this table no matter how long I struggled with it. To solve the problem, I loaded the data into Excel and created the following series of graphs:

Current World Opinions About the U.S.A

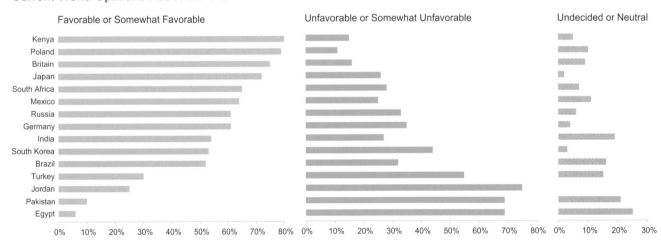

Source: 2004 study conducted by the Pew Research Center, as reported by the PBS television program NOW.

With this graphical display of the same information, the story suddenly came alive. It gave me the overview that I needed, plus some details that I never

FIGURE 1.4 This is a graphical display of the information that appears in the form of a table in *Figure 1.3.*

expected. For example, take a look at the country of Jordan, which is third up from the bottom. In addition to the fact that relatively few people in Jordan held favorable opinions of the U.S.A. (the green bar) and a greater percent than in any other country held unfavorable opinions (the red bar), notice that nobody in Jordan who was polled at that time was without an opinion, which pops out as an empty space where a gray bar is expected.

Here's another fairly typical example of a business graph that suffers from severe design problems:

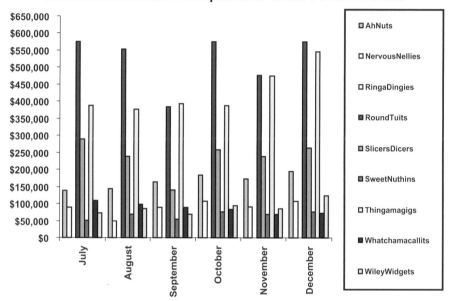

FIGURE 1.5 This is a typical example of a poorly designed vertical bar graph.

Without the graph's title, would you have any idea that its purpose is to compare the sales performance of the product named SlicersDicers to the performance of each of the other products? Designers speak of objects as having *affordances*—characteristics that reveal how they're supposed to be used and make them easy to use in those ways. A teapot has a handle. A door that you need to push has a push-plate. The design of an object should, in and of itself, suggest how the object should be used. This graph relies entirely on its title to declare its purpose. The design not only fails to suggest the graph's use, it actually subverts the user's ability to determine what the graph is for.

Imagine for a few minutes that it is your job to assist the manager responsible for the products that appear in this graph. You've been asked to create a simple means to help her compare the performance of SlicersDicers to the performance of the other products in her portfolio. For her immediate purposes, she isn't particularly interested in actual dollars. She just wants to know how recent sales of SlicersDicers compare to sales of her other products. To begin your task, examine the graph again; then, make a list of the ways its design fails to support the product manager's needs. Take a minute to write your list next to the graph in the right margin of the book.

For a thorough examination of *affordances* in the broader context of design for usability, see Donald A. Norman's classic text, *The Design of Everyday Things* (1988) Basic Books.

· · · · · · ·

Now, advance this exercise to the next level. Take a few more minutes to list the features you would incorporate into a new version of the graph to better serve the product manager's needs. Make use of your prior list to suggest areas that should be addressed.

. . . .

How did you do? Did you find yourself rising to the challenge? If so, you are already beginning to think critically and creatively about the design of tables and graphs. There is no single correct solution to the task at hand, but let me show one possible solution, which previews a design technique that may interest you.

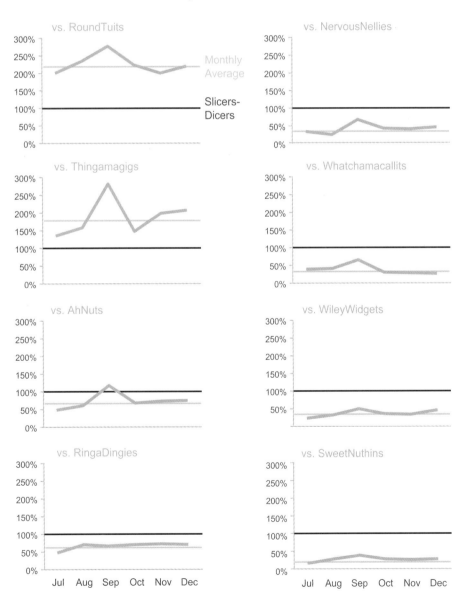

FIGURE 1.6 This is a series of related graphs, each designed to compare the sales of the SlicersDicers product to sales of a different product.

This solution uses a divide-and-conquer approach to the problem. With eight small graphs—one for each product that needs to be compared to SlicersDicers—the complications that resulted from squeezing everything into a single graph have been reduced. The black reference line that represents SlicersDicers' sales in each of the graphs serves as a powerful affordance, strongly suggesting that this display should be used to compare the sales performance of each product to SlicersDicers.

Let's look at one last example, for now, of ineffective graph design. Everyone is familiar with the ubiquitous pie chart. This one is quite typical in its design:

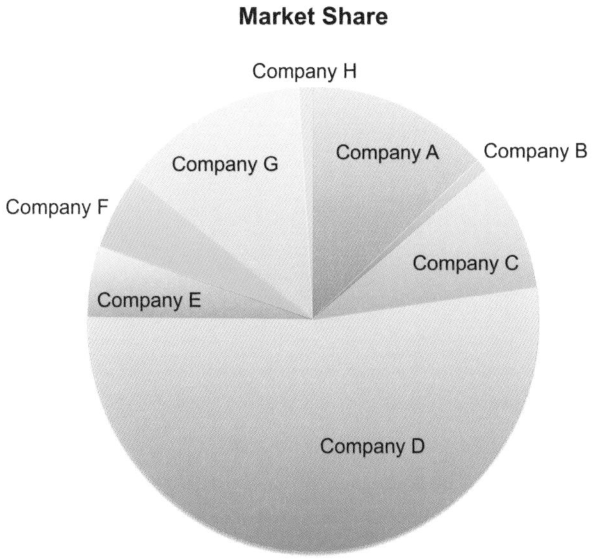

Market Share

FIGURE 1.7 This is an example of an ineffective style of graph: a pie chart.

Here's the same graph below, but this time it's dressed up with the addition of a 3-D effect:

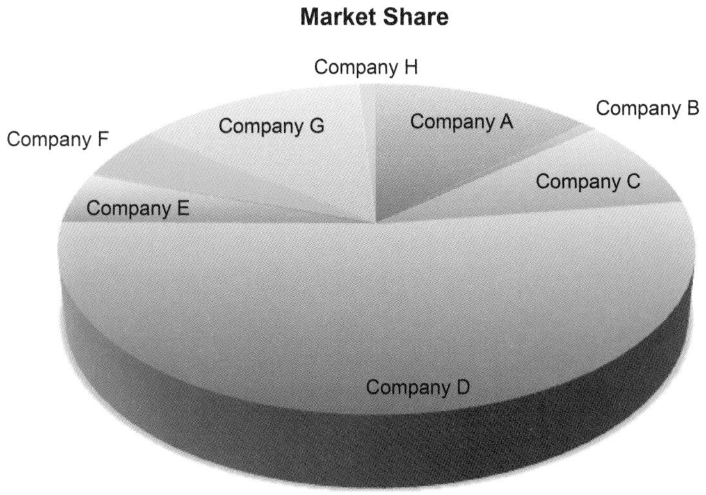

Market Share

FIGURE 1.8 This is a slight variation of *Figure 1.7*, which adds 3-D perspective. This design is even less effective.

Did the use of 3D enhance the display? These two graphs are meant to display the same market share numbers, yet notice how the addition of 3-D perspective affects your perception of the data. The intention of these graphs is for

Company G to compare its market share to the shares of its competitors. Can you determine Company G's market share from either of these graphs? Can you determine its rank compared to its competitors? Which has the greater share: Company A or Company G? Because of fundamental limitations in visual perception, you really can't answer any of these questions accurately.

Now look at the exact same market share values displayed more effectively:

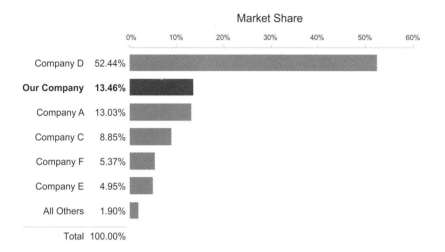

FIGURE 1.9 This horizontal bar graph displays the market share data in a way that is easy to interpret and compare.

Did you have any trouble interpreting this information? Did you struggle to find the most important information? It's obvious that Our Company ranks second, slightly better than Company A, and that our market share is precisely 13.46%. This display contains no distractions. It gives the numbers clear voices to tell their story.

Purpose

Although they are routinely used today, with few exceptions graphs have only been used to display quantitative information for a little more than two centuries. Even though tables have been around longer, it wasn't until the last quarter of the 20th century that the use of either became widespread. What caused this rise in their use? The personal computer.

Shortly after the advent of the PC, I began developing and teaching courses in the use of some of the earliest PC-based business software, including Lotus 1-2-3. Although Lotus wasn't the first electronic spreadsheet product, and it's no longer the most popular, it legitimized the PC as a viable tool for business. Prior to the advent of spreadsheet software, tables of quantitative information were generally produced using pencils, sheets of lined paper, a calculator, and hours of tedious labor. Graphs could only be produced using a pen (perhaps several, of different colors), a straight-edged device (e.g., a ruler or draftsman's triangle), a sheet of graph paper, and, once again, hours of meticulous labor. When chart-producing software hit the scene, many of us who would have never taken the time to draw a graph suddenly became Rembrandts of the X and Y axes , or so we thought. Like kids in a toy store, we went wild over all the available colors and cool effects, thrilled with the new means for techno-artistic expression. Through the *magic* of computers, making tables and graphs became easy— perhaps too easy.

Today, everyone can produce reports of quantitative information in the form of tables and graphs. Children in elementary school are taught the mechanics of using spreadsheet software. Something produced with a computer acquires an air of authenticity and quality that it doesn't necessarily deserve, however. In our excitement to produce what we could only do before with great effort, many of us have lost sight of the real purpose of quantitative displays: *to provide the reader with important, meaningful, and useful insight*. To communicate quantitative information effectively requires an understanding of the numbers and the ability to display their message for accurate and efficient interpretation by the reader.

The necessary knowledge and skills are well within your reach if you make a little effort. Once you've read this book and practiced a bit, you'll find that it is no more difficult and takes no longer to produce effective tables and graphs than it does to produce ineffective ones. By synthesizing the best practices of quantitative information design that have been learned through many years of research and real-world trial and error by trailblazers, I hope to make your effort relatively easy, and perhaps even enjoyable.

The purpose of quantitative tables and graphs is to communicate important information effectively. That's it. Not to entertain, not to indulge in self-expression, not to make numbers that you would otherwise find boring suddenly interesting through flash and dazzle. Edward Tufte, a well-known expert in this field, expresses this perspective quite simply: "The overwhelming fact of data graphics is that they stand or fall on their content, gracefully displayed . . . Above all else show the data." And to those who believe that they must dress up the numbers, he warns: "If the statistics are boring, then you've got the wrong numbers."[2]

As with any endeavor that's worthwhile, our goal in the design of quantitative information for the purpose of communication is *excellence*. Why would we aim for anything less when the means to attain excellence are readily available? Why anything less when doing good work is so satisfying, and simply getting by is such a drag? With the right knowledge and skills, we have a chance to make a difference. Even if you're not inspired by your organization's objectives, do it for yourself. Good work, or what Buddhists call "right livelihood," is one of funda-mental delights of life. It is fulfilling in and of itself.

As presenters of quantitative information, it is our responsibility to do more than sift through the information and pass it on; we must help our readers gain the insight contained therein. We must design the message in a way that leads readers on a journey of discovery, making sure that what's important is clearly seen and understood. The right numbers have important stories to tell. They rely on us to find those stories, understand them, and then tell them to others in a way that is clear, accurate, and compelling.

Scope

One of the challenges in writing a book like this is to constrain its scope. The contents must constitute a coherent whole that addresses a specific set of needs for a particular audience. There is a risk of communicating too little by saying

With few exceptions, the tables and graphs in this book were made using Microsoft Excel software, in part to demonstrate that good design can be achieved even if you have no software but Excel.

2. Edward R. Tufte (2001) *The Visual Display of Quantitative Information*, Second Edition. Graphics Press. The first quotation appears on pages 121 and 92; the second on page 80.

too much. This is indeed a challenge when your interests are far ranging, like mine. What to leave out? In light of this objective, I've limited this book to content that is 1) practical and applicable to typical organizational reporting, 2) focused on communication, and 3) focused on effective design.

Although some organizations employ professional statisticians to perform sophisticated quantitative analyses, most of us never touch advanced statistics and would be intimidated by anything much more complicated than a measure of average. Even though most organizations have a few specialized display techniques to fit their unique needs, display techniques that are not of general interest don't appear in this book. Rest assured that most or all of what you will find in this book will apply directly to your own work. You will not have to wade through information that might be interesting from an academic perspective but peripheral to the task at hand.

Tables and graphs of quantitative business information can be used for four purposes:

- Analyzing
- Communicating
- Monitoring
- Planning

When you use tables and graphs to discover the message in the data, you are performing analysis. When you use them to pass a message on to others, your purpose is communication. When you use them to track information about performance, such as the speed or quality of manufacturing, you are engaged in monitoring. When you use them to predict and prepare for the future, you are planning. All of these are important uses of tables and graphs, but the processes that you engage in and the design principles that you follow differ somewhat for each. My purpose in this book is to help you learn to design tables and graphs to communicate important information.

This book focuses on design. It is not about the mechanics of constructing tables and graphs using a particular software product (e.g., Microsoft Excel). It is not an encyclopedia of tables and graphs, listing and describing the countless variations that exist. It is not a book that will make you an expert in a particular type of graph (e.g., the nuances of histograms). It is not a book about making tables and graphs look pretty though there is certainly beauty in those that simply and elegantly hit their communication target spot on. It is a book about design, in particular about design practices that broadly apply to using tables and graphs for effective quantitative communication.

When do you use a table versus a graph? When do you use one type of graph rather than another? How do you highlight what's most important and make the message crystal clear? What should you avoid to ensure that you eliminate anything that might distract from your message? This book answers these questions and more, and the answers can be applied across the board to every table and graph you will ever need to create.

For instruction in the use of graphs for data exploration and analysis, see my book *Now You See It: Simple Visualization Techniques for Quantitative Analysis* (2009), and for instruction in the design of dashboards for performance monitoring, see my book *Information Dashboard Design: The Effective Visual Communication of Data* (2006).

For an excellent encyclopedia of charts, consider Robert L. Harris (1999) *Information Graphics: A Comprehensive Illustrative Reference*, Oxford University Press.

Intended Readers

Simply stated, this book is intended for anyone whose work involves the use of tables and graphs for presenting quantitative information. Those of us who hold this responsibility don't fit into a tidy collection of job titles. Our roles are spread across a broad spectrum. Some of us specialize in the production of analyses and reports, with job titles that often include the term analyst, for example: Financial Analyst, Business Analyst, Data Analyst, or Decision Support Analyst, to name a few. Some of us have managerial responsibilities and are occasionally required to prepare tables and graphs for other managers higher up in the ranks. Some of us are graphic artists, and someone decided that because the words graphs and graphic are related, responsibility for the production of quantitative graphs must belong to us. The rest of us are scattered over the organization. No matter what it says on your business card, if you are responsible for creating tables and graphs to communicate quantitative information, and you want to do it well, this book is for you.

Content Preview

Tables and graphs don't just display numbers; they present them in a manner that relates them to something, such as to time or to one another, to reveal a meaningful message in context. Tables and graphs are two members of a larger family of display methods known as *charts*. In addition to tables and graphs, there are other types of charts, such as diagrams, which illustrate a process or set of relationships, and maps, which depict information geospatially.

Although tables and graphs are both vehicles for presenting information visually, the role that visual perception plays in reading and interpreting the information presented differs significantly for tables versus graphs. Graphs are perceived almost entirely by our *visual* system, and, as such, employ a visual language of sorts. In a graph, lines, bars, and other objects positioned within a 2-D space formed by perpendicular axes are used to communicate visual patterns and relationships. To see patterns and relationships is a natural function of visual perception.

Tables, with their many columns and rows of text, interact primarily with our *verbal* system, which entails what we normally mean when we speak of language. As such, we process information contained in tables sequentially, reading down columns or across rows of numbers, comparing this number to that number, one pair at a time. This is different from the visual processing that occurs when we view graphs, which involves high-bandwidth, simultaneous input of more data, potentially enabling us to perceive a great deal of quantitative information in a burst of recognition. Neither method of quantitative display is inherently better than the other, but each is better than the other for particular communication tasks, and both play a vital role.

This book is designed to take you on a journey. We begin in Chapter 2, *Simple Statistics to Get You Started*, with an introduction to a few statistics that are easy to understand, incredibly handy, and particularly useful in tables and graphs.

We continue in Chapter 3, *Differing Roles of Tables and Graphs*, with an introduction to tables and graphs, including what's common to both, then proceed to which works best for particular purposes.

In Chapter 4, *Fundamental Variations of Tables*, we go on to break tables down into the types of information they can be used to display and into the basic ways they can be structured.

Before proceeding to a closer look at graphs, we take a brief detour in Chapter 5, *Visual Perception and Graphical Communication*, to learn how visual perception works, from the time when light enters our eyes to the time that the information that our brains have gleaned from that light is stored in memory for future reference or is immediately discarded. You may be tempted to skip this chapter. Don't give in. If you want to understand what works and what doesn't in the design of graphs so you can apply your knowledge to each new situation that arises, you will need to understand these basic concepts about visual perception.

In Chapter 6, *Fundamental Variations of Graphs*, we break graphs down into the types of information they can be used to display and explore the ways that data can be visually encoded to tell the story most effectively.

In Chapter 7, *General Design for Graphical Communication*, we apply what we've learned about visual design as general practices for both tables and graphs.

From there, we dive into the details of *Table Design* in Chapter 8. We examine the structural components of tables, how to combine them for optimal effect, and the all-too-common bad practices that you should avoid.

We then shift our focus in Chapter 9 to *General Graph Design*, beginning with design principles that apply to all types of graphs. In Chapter 10, *Component-Level Graph Design*, we look closely at each component of graphs to learn when and how to use them for effective communication. Chapter 11, *Displaying Many Variables at Once*, focuses on developing strategies for displaying complex messages. In Chapter 12, *Silly Graphs that Are Best Forsaken*, we learn to avoid graphs that don't present data effectively and what alternatives to use that do the job well.

In Chapter 13, *Telling Stories with Numbers*, we put the entire venture into perspective. Numbers are important because they have stories to tell that we must understand to make good decisions. We must learn to tell these stories in ways that are simple, clear, accurate, and compelling.

Finally, we end our journey in Chapter 14, *The Interplay of Standards and Innovation*, where I climb up on my soapbox and prophesy that you will suffer defeat if you don't establish and follow a practical set of standards for the design of tables and graphs.

Communication Style

At heart, I am a teacher. My mind works like a teacher's. When I tell you about something, I care that you get it. When you do, I feel good. When you don't, I feel that I've failed. Consequently, this book is designed as a learning experience, not simply to inform or entertain. It is not designed as a reference that you pull from the shelf occasionally. It is designed to get into your head in a way that is thorough and lasting.

As a result, this book is filled with examples that bring the material to life, as well as questions that invite you to think and perhaps even struggle a bit during the process. It contains exercises that provide an opportunity to practice what you're learning and to learn more thoroughly through that practice. It is laid out in a sequence that leads you through learning at a conceptual level, then allows you to apply your conceptual understanding to various real-world scenarios. I guide you on a journey of discovery, rather than presenting principles that you must memorize and follow based merely on my authority as an expert. I want you to learn these practices and make them a part of your work to the extent that you no longer need to think about them. If someone asks you why you design tables and graphs the way you do five years from now, I want you to still be able to explain it.

I love teaching, in part because I love learning. I try to approach each day as a student. Doing so enriches me and keeps life fresh and interesting. I invite you to approach this book with the curiosity of a student. If you do, your effort will be rewarded.

2   SIMPLE STATISTICS
TO GET YOU STARTED

Quantitative information forms the core of what organizations must know to operate effectively. The current emphasis on metrics, Key Performance Indicators (KPIs), Balanced Scorecards, and performance dashboards demonstrates the importance of numbers to organizations today. Stories contained in numbers can be communicated most effectively when we understand the fundamental characteristics and meanings of simple statistics that are routinely used to make sense of numbers, as well as the fundamental principles of effective communication that apply specifically to quantitative information.

Numbers are neither intrinsically boring nor interesting. The fact that they are quantitative in nature has no bearing on their inherent appeal. They simply belong to the class of information that communicates the quantity of something. The impact and appeal of information, quantitative or not, flow naturally from the significance and relevance of the message the information contains. As a communicator, it is up to you to give a clear and unhindered voice to that information and its story, using language that is easily understood by your audience.

You might be anxious to jump right into the design of tables and graphs. After all, that's the fun stuff. I must admit, I was tempted to get right to it, but because numbers are the substance of tables and graphs, it's important to begin our journey by getting acquainted with a few numbers that are particularly useful.

Quantitative Relationships

When you display quantitative information, whether you use a table or graph, the specific type of table or graph you use depends primarily on your story. What about the story? Quantitative stories are always about relationships. Numbers, in and of themselves, are of no use unless they measure something that's important. Here are some common examples of relationships that define the essential nature of quantitative stories:

Quantitative Information	Relationship
Units of a product sold per geographical region	Sales related to geography
Revenue by quarter	Revenue related to time
Expenses by department and month	Expenses related to organizational structure and time
A company's market share compared to that of its competitors	Market share related to companies
The number of employees who received each of the five possible performance ratings (1–5) during the last annual performance review	Employee counts related to performance ratings

In each of these examples, there is a simple relationship between some measure of quantity and one or more associated categories of interest (geography, time, etc.). Quantitative stories feature two types of data: *quantitative* and *categorical*. Quantitative values measure things; categories divide information into useful groups, and the items that make up each category identify the things that are measured. For example, geographical areas (e.g., north, east, south, and west) are items in a category that might be called sales regions and months are items in a category called time. This distinction between quantitative values and categorical items is fundamental to tables and graphs. Quantitative values and categorical items serve different but complementary purposes and are often structured and displayed in distinct ways.

Sometimes the quantitative relationships we display are simple associations between quantitative values and the categorical items that label them, such as those in the previous examples. Sometimes the relationships display direct associations between different sets of quantitative values, such as the number of marketing emails that were sent in relation to the resulting number of orders received, or the percentage of times that doctors forget to wash their hands in hospitals in relation to the percentage of infections. This distinction between simple relationships that associate quantitative values and categorical items and somewhat more complex relationships that associate multiple sets of quantitative values is also fundamental to our use of tables and graphs. Different types of relationships require different types of displays.

So far we've only examined a few examples, but the list of potential quantitative relationships is endless. Think for a minute or two about the quantitative information that's important to your organization. Can you think of any information that doesn't involve relationships?

Quantitative values are expressed in units of measure. For instance, the quantitative value $200 is made up of the quantity—200—and a relevant unit of measure—dollars.

Thus far we've learned the following about quantitative stories:

- Quantitative stories include two types of data:
  - Quantitative
  - Categorical

- Quantitative stories always feature relationships.
- These relationships involve either:
  - Simple associations between quantitative values and corresponding categorical items or
  - More complex associations among multiple sets of quantitative values.

In addition to the two fundamental types of quantitative relationships that we've already noted, there are also a variety of ways in which categorical items or the quantitative values associated with them can relate to one another. Let's take a look.

Relationships Within Categories

Categorical items that we use in tables and graphs to label corresponding measures can relate to one another in the following ways:

- Nominal
- Ordinal
- Interval
- Hierarchical

NOMINAL

A *nominal* relationship is one in which the values in a single category are discrete and have no intrinsic order. For instance, the four sales regions East, West, North, and South have no particular proper order in and of themselves. These labels simply name the different sales regions, thus the term "nominal," which means "in name only." Here's an example:

Region	Sales
North	139,883
East	135,334
South	113,939
West	188,334
Total	$577,490

FIGURE 2.1 This is an example of a nominal relationship.

When you tell a quantitative story that is nominal in nature, you associate the quantitative values with the corresponding categorical labels, but your story does not relate the categorical items to one another in any particular way.

ORDINAL

In an *ordinal* relationship, the categorical items have a prescribed *order*. Typical examples include "first, second, third..."; "small, medium, and large"; and "best salesperson, second best salesperson...". To display them in any other order, except in reverse, would rarely be meaningful.

INTERVAL

An *interval* relationship is one in which the categorical items consist of a sequential series of numerical ranges that subdivide a larger range of quantitative values into smaller ranges. These smaller numerical ranges, called intervals, are arranged in order from smallest to largest. Here's a typical example:

Order Size (U.S. $)	Order Quantity	Order Revenue
>= 0 and < 1,000	17,303	6,688,467
>= 1,000 and < 2,000	15,393	26,117,231
>= 2,000 and < 3,000	10,399	29,032,883
>= 3,000 and < 4,000	2,093	6,922,416
>= 4,000 and < 5,000	1,364	5,805,184
Total	46,552	$204,515,383

FIGURE 2.2 This is an example of an interval relationship. Notice that Order Size consists of a sequential series of numerical ranges that subdivide a larger range of quantitative values.

In this example, to see how the orders were distributed across the entire range of order sizes, it wouldn't make sense to count the number of orders and sum their totals for each individual order amount because that would involve an unmanageably large set of order sizes. The solution involves subdividing the full range of order sizes into a series of sequential, equally sized intervals.

Take a moment to test what you've learned so far. Look at the example below and determine which of the three relationships—nominal, ordinal, or interval—best describes its categorical items of time (months in this case).

Department	Jan	Feb	Mar	Q1 Total
Marketing	83,833	93,883	95,939	273,655
Sales	38,838	39,848	39,488	118,174
HR	37,463	37,939	37,483	112,885
Finance	13,303	14,303	15,303	42,909
Total	$173,437	$185,973	$188,213	$547,613

FIGURE 2.3 This is an example of time-series relationship.

Your initial inclination was probably to call this an ordinal relationship, for months usually make sense only when arranged chronologically. However, items that make up an interval scale also have a proper order, which invites the question: "Do these units of time represent intervals along a quantitative scale? The answer is "Yes, they do." Time is a quantitative scale that measures duration. Even though the months in the example above do not all represent the same exact number of days and are therefore not equally sized intervals, for reporting purposes we treat them as equal.

So far the categorical relationships that we've examined involve relationships between items in the same category. The remaining relationship, discussed next, is different.

HIERARCHICAL

A *hierarchical* relationship involves multiple categories that are closely associated with one another as separate levels in a series of "parent-to-child" connections. If we start from the top of the hierarchy and progress downward, each item at each level is associated with only one item at the level above it. Each item at every level, except the bottom level, however, can have one or more items associated with it in the next level down. This is much easier to show than to describe with words. Here's a typical example, viewed from left to right:

Division	Dept	Group	Expenses ($)
G&A	Human Resources	Recruiting	42,292
		Compensation	118,174
	Info Systems	Operations	512,885
		Applications	442,909
Finance	Accounting	AP	73,302
		AR	83,392
	Corp Finance	Fin Planning	93,027
		Fin Reporting	74,383

FIGURE 2.4 This is an example of a hierarchical relationship. The G&A division is composed of two departments: Human Resources and Info Systems. The Recruiting and Compensation groups belong to the Human Resources department, and the Operations and Applications groups belong to the Info Systems department.

Hierarchical relationships between categories are used routinely in tables to organize quantitative information.

Relationships Between Quantities

Categorical items can also relate to one another by virtue of the quantitative values associated with them. The quantitative values can be arranged to display the following relationships:

- Ranking
- Ratio
- Correlation

RANKING

When the order in which the categorical items are displayed is based on the associated quantitative values, either in ascending or descending order, the relationship is called a *ranking*. If you need to construct a list of your company's top five sales orders for the current quarter based on revenue, the story would be enhanced if you arranged them by size, in this case from the largest to the smallest, as you see in the following figure:

Rank	Order Number	Order Amount
1	100303	1,939,393
2	100374	875,203
3	100482	99,303
4	100310	87,393
5	100398	67,939
		$3,069,231

Technically, the term ordinal could be used to describe a ranking relationship as well, but I'm using distinct terms to highlight the difference between a sequence based on categorical items and one based on quantitative values.

FIGURE 2.5 This is an example of a ranking relationship.

RATIO

A *ratio* is a relationship that compares two quantitative values by dividing one by the other. This produces a number that expresses their relative quantities. A common example is the relationship of the quantitative value for a single categorical item compared to the sum of the entire set of values in the category (e.g., the sales of one region compared to total sales of all regions). The ratio of a part to its whole is generally expressed as a percentage where the whole equals 100%, and the part equals some lesser percentage. Here's an example of a part-to-whole relationship that displays market share information for five companies, both in actual dollar sales and in percent-of-total sales:

Company	Sales	Sales %
Company A	239,949,993	15%
Company B	873,777,473	54%
Company C	37,736,336	2%
Company D	63,874,773	4%
Company E	399,399,948	24%
Total	$1,614,738,523	100%

FIGURE 2.6: This is an example of a part-to-whole ratio.

When you want to compare the size of one part to another or to the whole, it is easier, more to the point, and certainly more efficient for your audience to interpret a table or graph that contains values that have been expressed as percentages. This is true because percentages provide a common denominator and common frame of reference, and not just any common denominator but one with the nice round value of 100, which makes comparisons easy to understand.

Another common use of ratios involves measures of change. When the value of something is tracked through time, it is often useful to note how it changes from one point in time to the next. Here's an example of a ratio that expresses the degree of change, in this case change in expenses from one month to the next:

Department	Expenses Jan	Feb	Change	Change %
Sales	9,933	9,293	-640	-6%
Marketing	5,385	5,832	+447	+8%
Operations	8,375	7,937	-438	-5%
Total	$23,693	$23,062	-$631	-3%

FIGURE 2.7 This is an example of a ratio used to compare the expenses from one month to the next.

CORRELATION

A *correlation* compares two paired sets of quantitative values to determine whether increases in one correspond to either increases or decreases in the other. For instance, is there a correlation between the number of years employees have been doing particular jobs and their productivity in those jobs? Does productivity increase along with tenure, does it decrease, or is there no significant correlation in either direction? Correlations are important to understand, in part because they make it possible to predict what will happen to values in one variable (e.g., sales revenue) by knowing or perhaps even controlling values in another variable (e.g., marketing emails).

Thus far we've learned the following about quantitative information:

- Quantitative stories include two types of data:
  - Quantitative
  - Categorical
- Quantitative stories always feature relationships.
- These relationships involve either:
  - Simple associations between quantitative values and corresponding categorical items or
  - More complex associations among multiple sets of quantitative values
- Categorical items exhibit four types of relationships:
  - Nominal
  - Ordinal
  - Interval
  - Hierarchical

- Quantitative values exhibit three types of relationships:
  - Ranking
  - Ratio
  - Correlation

We have not covered a comprehensive list of possible quantitative relationships. Rather, we've considered only those that are most relevant when presenting quantitative information in typical ways. If you're wondering why these different quantitative relationships are important enough to cover in this chapter, hold on for a while. When we get to later chapters on tables and graphs, the importance of these relationships and your ability to identify them will become clear. You'll discover that there are many specific ways to design tables and graphs that tie directly to these specific quantitative relationships.

Numbers that Summarize

Statistics provide several ways reduce or summarize data. Summarization is also referred to as *aggregation*. Often, your quantitative message is best communicated by reducing large sets of numbers to a few numbers, allowing your readers to easily and efficiently comprehend and assimilate the story. If an executive asks you how sales are doing this quarter, you wouldn't give her a report that listed each individual sales order; you would give her the information in summary form. Relevant data might include such aggregates as the sum of sales orders in U.S. dollars, the count of sales orders, and perhaps even the average sales order size in U.S. dollars.

We have several ways to summarize numbers, some that are visual in nature and apply only to graphs, which we'll explore more thoroughly later, and some that are purely statistical in nature, which we'll examine now. Summing and counting sets of numbers are the most common aggregations used in quantitative communication. I assume that you already understand counts and sums, so we'll skip them and proceed directly to other less familiar data reduction methods that are also useful.

Measures of Average

Let's begin by examining what we already know about averages. Take a moment to finish this sentence: "An average represents . . ."

· · · · · · ·

It's interesting how many terms we carry around in our heads and use without really knowing how to define them. Ever had a child ask you to explain something quite familiar and found yourself struggling for words? If the concept of an average is one of those terms for you, here's a definition:

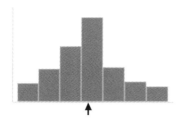

> An average is a single number that represents the center of an entire set of numbers.

There are actually four distinct ways in statistics to measure the center of a set of numbers, and all of them are called averages:

- Mean
- Median
- Mode
- Midrange

It's useful to understand how these four differ. Selecting the wrong type of average for your message could mislead your audience.

MEAN

Normally, when most of us think of an average, what we have in mind is more precisely called the *arithmetic mean* or simply the *mean*. In fact, many software products label the function that calculates the mean as "average" (or sometimes "AVG"). Statisticians must cringe when they see this. Statistical software wouldn't make this mistake. Means are calculated as follows:

> Sum all the values and then divide the result by the number of values.

Here's an example:

Quarter	Units Sold
Q1	339
Q2	373
Q3	437
Q4	563
Sum	1,712
Count	4
Mean (per Qtr)	428

FIGURE 2.8 This is an example of a mean, calculated as 1,712 (the sum) divided by 4 (the count), equaling 428.

A mean is a simple form of average to calculate, but isn't always the best choice for your story.

Means measure the center of a set of numbers in a way that takes every value into account, no matter how extreme. Sometimes this is exactly what you need, but sometimes not. Take a look at the following example, and see if you can determine why using the mean would produce a misleading summary of employees' salaries in the marketing department if your objective is to express the typical salary.

Employee	Position	Annual Salary
Employee A	Vice President	475,000
Employee B	Manager	165,000
Employee C	Manager	165,000
Employee D	Admin Assistant	43,000
Employee E	Admin Assistant	39,000
Employee F	Analyst	65,000
Employee G	Analyst	63,000
Employee H	Writer	54,000
Employee I	Writer	52,000
Employee J	Graphic Artist	64,000
Employee K	Graphic Artist	62,000
Employee L	Intern	28,000
Employee M	Intern	25,000
	Mean Salary	$100,000

FIGURE 2.9 This is an example of the use of a statistical mean in circumstances for which it is not well suited.

Why doesn't the mean work well for this purpose? In this case the mean is much higher than most salaries, giving the impression that employees are better compensated than they typically are. What you're seeing here is the fact that the mean is very sensitive to extremes. The Vice President's salary is definitely an extreme, a value far above the norm. When you need a measure of center that represents what is typical of a set of values, you should use an average that isn't sensitive to extremes.

Statisticians refer to extreme values in a data set (i.e., those that are located far away from most of the values) as outliers. The Vice President's salary in *Figure 2.9* is an outlier.

MEDIAN

The *median* is an expression of average that comes in handy when you need to tell quantitative stories such as the one in the previous example because the median is not at all sensitive to extremes and therefore does a better job of expressing what's typical.

Medians are calculated as follows:

> Sort the values in order (either high to low or low to high) and then find the value that falls in the middle of the set.

Here are the same salaries, but this time sorted from high to low so we can easily determine the median:

If you are using software or a calculator that supports the calculation of the median, you won't need to sort the set of numbers and manually select the middle value.

Rank	Position	Annual Salary
1	Vice President	475,000
2	Manager	165,000
3	Manager	165,000
4	Analyst	65,000
5	Graphic Artist	64,000
6	Analyst	63,000
7	Graphic Artist	62,000
8	Writer	54,000
9	Writer	52,000
10	Admin Assistant	43,000
11	Admin Assistant	39,000
12	Intern	28,000
13	Intern	25,000
	Median Salary	$62,000

FIGURE 2.10 This is an example of the use of the statistical median.

This data set contains 13 values, so the value that resides precisely in the middle is the seventh, which is $62,000. If you want to communicate the typical marketing department salary, $62,000 would clearly do a better job than $100,000. However, if your purpose is to summarize the salaries of each department in the company to show their comparative impact on expenses, which type of average would work better: the median or the mean? In this case the mean would be the better choice because you want a number that fully takes all values into account, including the extremes. To ignore them through use of the median would undervalue financial impact.

The median is actually an example of a special kind of value called a *percentile*. A percentile expresses the percentage of values that fall below a particular value. The median is just another name for the 50th percentile, for it expresses the value on or below which 50% of the values fall.

You might have noticed while considering how to determine the median above that I ignored a potential complication in the process. What do you do if your data set contains an even number of values, rather than an odd number like the 13 employee salaries above? In this case, you simply take the two values that fall in the middle of the set (e.g., the fifth and sixth values in a set of ten) and then determine the value that's halfway between the two. In fact, you can use the same method that you use for calculating the mean to find the value halfway between the two middle values: sum the two middle values then divide the result by two. If you're using software or a calculator to determine the median, this process is handled for you automatically.

MODE AND MIDRANGE

The two remaining types of averages—modes and midranges—are rarely useful to non-statisticians, but let's take a moment to understand them anyway.

The *mode* is simply the specific value that appears most often in a set of values. In the set of marketing department salaries that we examined previously, the mode is $165,000 because this is the only value that appears more than once in the set. As you can see, the mode wouldn't be a useful means of expressing the center of the marketing department's salaries. The most common value in a data set, especially a small one, isn't necessarily anywhere near the center. If no value appears more than once, the set doesn't even have a mode. If two values appear twice in the set and no other values appear more than once, the set is bimodal. If more than two values appear more than once with the same high degree of frequency, the set is *multimodal*. Modes are rarely useful for most data presentation purposes.

The final method for expressing the center of a set of values is the simplest to calculate, but you get what you pay for. It's called the *midrange*. The midrange is the value midway between the highest and lowest values in a set of values. To calculate the midrange, you find the highest and lowest values in the set, add them together, and then divide the result by two. This method is an extremely fast way to calculate an average. If you're on the spot for a quick estimate, you can use the midrange, but be careful. Unless the values are uniformly distributed across the range, the midrange is far too sensitive to the extremes of the highest and lowest values. You're better off using the mean or the median.

Measures of Variation

It is often useful to present more than the center of a set of values. For example, sometimes you need to communicate the degree to which values vary, such as the full range across which the values are distributed. Two sets of values can have exactly the same average, but one set could be distributed across a broad range while the other is tightly grouped around its average. In some cases this difference is significant. Values that vary widely are volatile. Perhaps they shouldn't be, so you're helping your organization by pointing this out. For example, if salaries for the same basic position vary greatly within your organization, this may be a problem worth noting and correcting. It might also be useful to recognize and communicate to senior management that sales in January for the past 10 years were always only 4% of annual sales, varying no

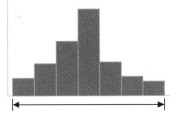

more than half a percent either way from year to year. Such a pattern, with no significant variation, despite expensive marketing campaigns, might indicate that the marketing budget should be saved for later in the year. Values that fall far outside the normal range might indicate underlying problems or even extraordinary successes that should be investigated. A salesperson with an unusually high order-return ratio might be selling products to his customers that they don't need. A department with exceptionally low expenses per employee might have something useful to share with the rest of the company.

Variation in a set of values can be expressed succinctly through the use of a single number, but there are multiple methods. We will examine the two that are typically most useful:

- Spread
- Standard Deviation

Like averages, these two measures of variation each work best in specific circumstances. Let's use an example consisting of two sets of values to illustrate these circumstances. Imagine that you work for a manufacturer that uses two warehouses to handle the storage of inventory and the shipping of orders. You've been receiving complaints from customers about shipments from Warehouse B. To simplify the example, let's say that you've gathered information from each warehouse about shipments of 12 orders of the same product during the same period of time. Ordinarily you would gather shipment information for a much larger set of orders to ensure a statistically significant sample, but we'll stick with a small data set to keep the example simple. Here are the relevant values, which in this case are the number of days it took for each of the 12 orders to be processed, from the day each order was received to the day it was shipped:

| | Days to Ship | |
Order	Warehouse A	Warehouse B
1	3	1
2	3	1
3	3	1
4	4	3
5	4	3
6	4	4
7	5	5
8	5	5
9	5	5
10	5	6
11	5	7
12	5	10

FIGURE 2.11 This table shows the days it took to ship two sets of 12 orders, one set from Warehouse A and one from Warehouse B.

Because the use of sums and averages is such a common way of analyzing and summarizing quantitative information, you could begin by performing these calculations, resulting in the following:

Warehouse	Sum	Mean	Median
A	51	4.25	4.5
B	51	4.25	4.5

FIGURE 2.12 This table contains various values that summarize the number of days it took the two warehouses to each ship a set of 12 orders.

If you were locked into this one way of summarizing and comparing sets of numbers, you might conclude and consequently communicate that the service provided by Warehouse B is equal to that of Warehouse A. If you did, you would be wrong.

The significant difference in performance between the two warehouses jumps out at you when you focus on the variation. Warehouse A provides a consistent level of service, always shipping orders in three to five days from the date they're received. Shipments from Warehouse B, however, are all over the map. Sometimes it fulfills orders much faster than Warehouse A, and at others times its performance is much slower. It's likely that the complaints came from customers who received their orders after waiting longer than five days and perhaps also from regular customers who, like most, value consistency in service, and find it annoying to receive their orders anywhere from one to ten days after placing them. Given this message about the inconsistent performance of Warehouse B, let's take a look at the two available ways to measure and communicate this variation.

SPREAD

The simpler of the two methods is called the *spread*. You can calculate the spread as follows:

> Subtract the lowest value from the highest value.

That's it. This is a measure of variation that everyone can understand, which is its strength. To summarize variation in the performance of Warehouse A versus Warehouse B, you could do so as follows:

	Warehouse A	Warehouse B
Range of days to ship	2	9

FIGURE 2.13 This table shows the ranges of days it took the warehouses to ship the two sets of orders.

Similar to the midrange averaging method, the spread method of describing variation suffers from its reliance on too little information (only the highest and lowest values), which robs it of the greater accuracy and usefulness of the standard deviation method that we'll examine next. The spread also suffers from the fact that it is very affected by extreme values. If Warehouse B had shipped seven orders in 5 days, one order in 1 day, and one order in 10 days, that would be a different story from the one contained in the data, but the spread would be the same. Despite its limitations, spread is a characteristic of variation that's useful to know.

STANDARD DEVIATION

The single measure of variation that reveals more than others is the *standard deviation*. Here's a definition:

> The standard deviation measures variation in a set of values
> relative to the mean.

The bigger the standard deviation, the greater the range of variation relative to the mean. This becomes a little clearer when you visualize it. Look at the

number of days it took Warehouse B to ship each order compared to the mean
value of 4.25 days:

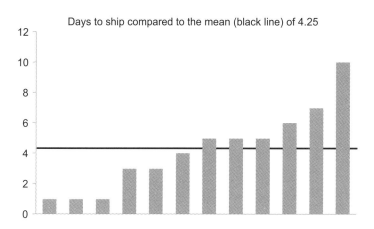

FIGURE 2.14 This graph shows a
simple way to visualize the days it
took Warehouse B to ship each of the
12 orders compared to the mean
value of 4.25 days.

Better yet, because our purpose here is to examine the degree to which the
shipments of the individual orders varied in relation to the mean, this graph
makes it a little easier to visualize:

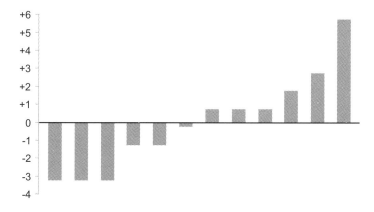

FIGURE 2.15 This graph displays the
days it took Warehouse B to ship the
individual orders relative to the
mean.

So far we haven't displayed the standard deviation. We're still leading up to
that. The standard deviation will provide a single value that summarizes the
degree to which the 12 shipments as a whole varied in relation to the mean (i.e.,
average degree of variation). Standard deviation can be determined as follows:

1. Calculate the mean of the set of values.
2. Subtract each individual value in the set from the mean, resulting in a list
   of values that represent the differences of the individual values from the
   mean.
3. Square each of the values calculated in step 2.
4. Sum the values calculated in the step 3.
5. Divide the value calculated in step 4 by the total number of values.
6. Calculate the square root of the result from step 5.

Looks like a lot of work, doesn't it? To complicate matters further, there are technically two formulas for calculating a standard deviation, one for the standard deviation of an entire population of values, and one for the standard deviation of a sample set of values. The steps above are used for an entire population of values. If the set consists only of a sample of the entire population of values, step 5 above would differ in that you would divide by the number of values minus 1, rather than simply by the number of values. Fortunately, most software products that produce tables and graphs include a simple way to calculate the standard deviation, so you don't have to perform the calculations yourself.

Because the set of values that measure the number of days it takes for Warehouse B to ship orders is only a sample set of values (i.e., 12 orders that shipped on one particular day), we'll use the form of the calculation that's used for sample sets, which produces a standard deviation of 2.70 days. We can compare this to the standard deviation for Warehouse A's shipments of 0.86 days. The difference between a standard deviation of 2.70 and one of 0.86 indicates a much higher degree of variation in Warehouse B's shipping performance compared to Warehouse A's. Standard deviations are a concise measure that can be used to compare relative variation among multiple sets of values.

In addition to its use for comparisons, a single standard deviation can also tell you something about the degree to which the values vary. However, the ability to look at a standard deviation and interpret the range of variation that it represents requires a little more knowledge.

In general, when individual instances of almost any type of event are measured and those measurements are arranged by value from lowest to highest, most values tend to fall somewhere near the center (e.g., the mean). The farther you get from the center, the fewer instances you will find. If you display this in the form of a graph called a *frequency polygon*, which uses a line to trace the frequency of instances that occur for each value from lowest to highest, you have something that looks like a bell-shaped curve, called a *normal distribution* in statistics.

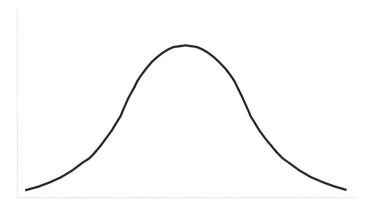

FIGURE 2.16 This curved line represents a normal distribution. It displays the frequency of values as they occur from the lowest value at the left to the highest value at the right. Most instances have values near the midpoint of the set of values, which represents the mean. In a perfect normal distribution, the frequency of instances decreases at the same rate to the left and to the right of the mean, resulting a curve (i.e., the black line) that is symmetrical.

So how do normal distributions relate to our examination of standard deviations? When you have a normal distribution, the standard deviation describes variation as percentages of the whole. The following figure overlays the normal distribution displayed in the previous figure with useful information that the standard deviation reveals.

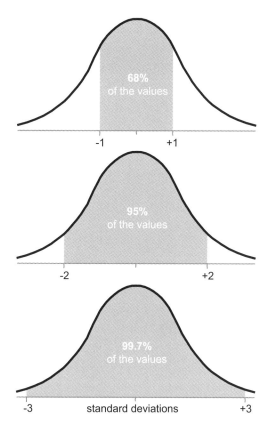

FIGURE 2.17 This figure shows a normal distribution of values in relation to the standard deviations of those values. The percentages of values that fall within one, two, and three standard deviations from the mean can be predicted with a normal distribution, and consequently can be predicted to a fair degree with anything that is close to a normal distribution. This is called the empirical rule.

With normal distributions, 68% of the values fall within one standard deviation above and below the mean, 95% fall within two standard deviations, and 99.7% fall within three. Stated differently, if you're dealing with a distribution that's close to normal, you automatically know that one standard deviation from the mean represents approximately 68% of the values, and so on. Given this knowledge, the standard deviation of a set of values has meaning in and of itself, not just as a tool for comparing the degree of variation among two or more sets of values. The bigger the standard deviation, the broader the range of values, and thus the greater difference in variation between them.

How does this relate to your world and the types of phenomena that you examine and present? Take a couple of minutes to list a few examples that are good candidates for measures of variation. In what situations would specific degrees of variation indicate something important to your organization?

. . . .

Here are a few examples that I've encountered:

- Variation in the selling price of specific products or services. Is variation greater in some parts of the world or for some sales representatives? Do differences in variation correspond to increased or decreased profits?

- Different degrees of variation in measures of performance, such as the time it takes to manufacture products, answer phone calls, or resolve technical problems. Do instances of greater variation indicate problems in training, employee morale, process design, or systems? What does a greater degree of variation today compared to the past signify?
- Variation in employee compensation. Why is there such a discrepancy in compensation for the same job in different departments? Does this broad variation in salaries have an effect on employee morale or performance?
- Variation in the cost of goods purchased from different vendors. Why is variation in costs associated with some vendors so much more than others for the same goods?
- Variation in departmental expenses. How is it that some departments manage to keep their expenses so much lower than other departments?

I could go on, but I think the point is clear. Measures of variation tell important stories, so knowing ways to summarize and concisely communicate these stories is indeed useful.

Measures of Correlation

Earlier in this chapter, I described correlation as a particular type of quantitative relationship where two paired sets of quantitative values are compared to one another to see if they correspond (i.e., co-relate) in some manner. For instance, does tenure on the job relate to productivity? In this section we're going to look at a particular way to measure correlation and express it as a single value. This single value is called the *linear correlation coefficient*. It answers the following questions:

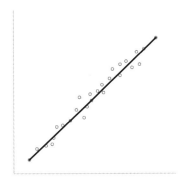

- Does a correlation exist?
- If so, is it strong or weak?
- If so, is it positive or negative?

Here's a concise definition:

> The linear correlation coefficient measures the direction (positive or negative) and degree (strong or weak) of the linear relationship between two paired sets of quantitative values.

By "two paired sets of quantitative values" I mean the two sets of values that are considered when you examine the relationship of one measurable thing (a.k.a., a variable) to another, such as an employee's tenure (e.g., number of years on the job) to his productivity on the job (e.g., number of items manufactured per hour). In this case, the number of years on the job and items manufactured per hour for all employees constitute a paired set of values. By "linear correla-tion" I mean a consistent relationship between two things, for instance, if you measure the correlation between employee tenure and productivity and find

A variable is something that can have multiple values, such as employee productivity.

that as tenure increases, productivity also increases, or that as tenure increases, productivity actually decreases. A linear correlation is limited, however, in that it cannot describe a relationship that is inconsistent, for example, if productivity increases along with tenure to a point but after that point it decreases as tenure continues to increase. This is still a relationship, but it's nonlinear. The direction of a correlation is either positive or negative. With positive correlations between two sets of values (A and B), as the value of A increases, the value of B likewise increases, and as the value A decreases, so does the value of B. With negative correlations, as the value of A increases, the value of B decreases, and vice versa.

If you had to calculate the linear correlation coefficient manually, you'd have to work through several steps. Very few of us need to do this because software or calculators do this for us. What really matters is that we know how to interpret the resulting value, so let's focus on the number itself and what it means.

Despite its intimidating name, the linear correlation coefficient is actually quite simple to interpret. Here are a few guidelines:

- All values fall between +1 and -1.
- A value of 0 indicates that there is no linear correlation.
- A value of +1 indicates that there is a perfect positive linear correlation.
- A value of -1 indicates that there is a perfect negative linear correlation.
- The greater the value, either positive or negative, the stronger the linear correlation.

It's getting clearer, but it will still help to look at this visually. To do so, we're going to use a graph called a *scatter plot*, which is specifically designed to display the correlation of two paired sets of quantitative values. Perhaps you've seen this type of graph listed as one that's available in software but have never used it, or perhaps you have only a vague idea how it works. With a little exposure, you'll find that scatter plots are quite easy to use and interpret and quite useful for revealing and communicating quantitative relationships.

Here's a series of scatter plots that will help you visualize the types of relationships that a linear correlation coefficient is designed to reveal. Each graph displays the relationship between two paired sets of values, one horizontally along the X axis and one vertically along the Y axis. When you read a scatter plot, you should look for what happens to the value along the Y axis in relation to what happens to the value along the X axis. As X goes up, what happens to Y? As X goes down, what happens to Y? Is the relationship strong (i.e., it's close to a straight line) or is it weak (i.e., it bounces around)? Is it positive (i.e., it moves upward from left to right) or is it negative (i.e., it moves downward from left to right)? Each of the following graphs displays a different relationship between the variable plotted along the X axis (horizontal) and the variable plotted along the Y axis (vertical), with the linear correlation coefficient in parentheses to help you understand its meaning.

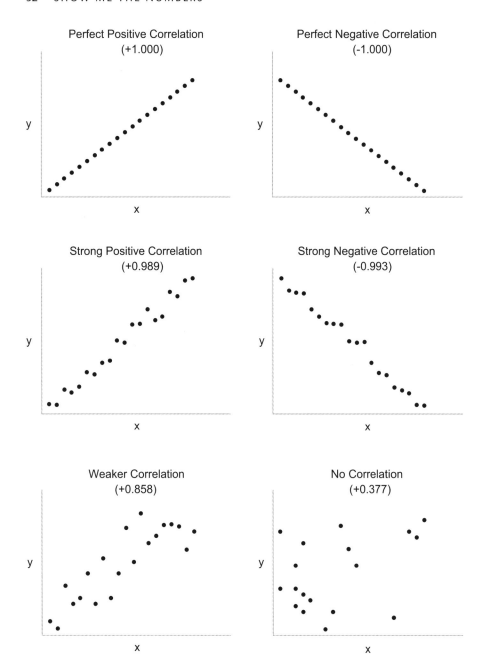

FIGURE 2.18 This is a series of scatter plots, each displaying a different relationship between two sets of paired values (e.g., employee tenure and productivity).

Bear in mind that these scatter plots are simply examples of correlations. If the linear correlation coefficient in the left-middle scatter plot were +0.970 rather than +0.989, it would still represent a strong positive relationship.

One way of looking at linear correlations as displayed in scatter plots is to imagine a straight line that passes through the center of the dots; then, determine the strength of the correlation based on the degree to which the dots are tightly grouped around that line: the tighter the grouping, the stronger the relationship. Here are examples of how scatter plots would look if you actually drew the lines:

Drawing a straight line of best fit through the center of the series of points in a scatter plot is a common technique for highlighting the relationship between two sets of values. It's called a *linear trend line*, *straight line of best fit*, or, more formally, a *regression line*.

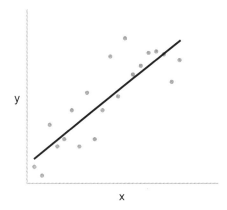
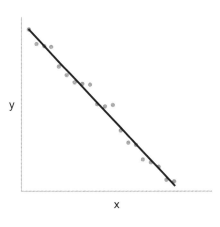

FIGURE 2.19 These are scatter plots with lines of best fit through the center of the dots to clearly delineate the nature of the relationship.

Based on what you've learned about scatter plots, how would you describe each of the relationships displayed above?

. . . .

In the scatter plot on the left, the characteristics you must consider are:

- The direction of the line, which in this case is upward from left to right
- The closeness of the grouping of dots around the line, which in this case is not particularly tight

Given these two observations, we can say that the scatter plot on the left depicts a correlation that is positive (i.e., upward from left to right) but not extremely strong (i.e., not tightly grouped around the line). Using this same method of interpretation, the scatter plot on the right depicts a correlation that is negative and very strong but not perfectly so.

At this point, you may be wondering: "At what value of a linear correlation coefficient does a correlation cease to be strong and begin to become weak or cease to be a correlation at all?" There is no precise answer to this question. It depends to some degree on the number of paired values included in your data sets; the more values you have, the greater confidence you can have in the validity of the linear correlation coefficient. Because our purpose here is not to delve too deeply into the realm of statistics, let's be content with the knowledge that values close to 1 in positive correlations and close to -1 in negative correlations indicate strong relationships, and that the closer they are to 1 or -1, the stronger the relationship.

Remember, linear correlation coefficients can only describe relationships that are linear—that is, ones that move in one direction or another roughly in a straight line—but not relationships that are positive under some circumstances and negative under others. Here's such an example:

For an excellent introduction to statistics, including much more information than I've provided about correlations, I recommend the textbook by Mario F. Triola (2009) *Elementary Statistics*, Eleventh Edition. Addison Wesley Longman Inc.

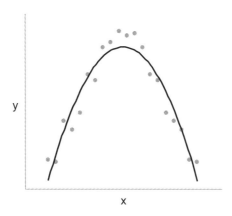

FIGURE 2.20 This is an example of a nonlinear correlation.

What you see here is definitely a correlation of sorts, but it certainly isn't linear (can't be described by a straight line). If this scatter plot represents the relationship between employee tenure (i.e., years on the job) on the X axis and employee productivity on the Y axis, how would you interpret this relationship, and how might you explain what is happening to productivity after employees reach a certain point in their tenure?

· · · · · · · ·

After studying this scatter plot and double-checking the data, you would likely suggest that something be done as employees reach the halfway point along their tenure timelines, such as offering new incentives to keep them motivated or retraining them for new positions that they might find more interesting.

Measures of Ratio

In contrast to correlations, which measure the relationship between multiple paired sets of values, a ratio measures the relationship between a single pair of values. A typical example that we encounter in business is the book-to-bill rate, which is a comparison between the value associated with sales orders that have been booked (i.e., placed by customers and accepted as viable orders) and the value associated with actual billings that have been generated in response to orders.

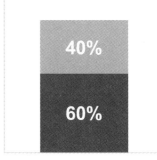

Ratios can be expressed in four ways:

- As a sentence, such as "Two out of every five customers who access our website place an order."
- As a fraction, such as 2/5 (i.e., 2 divided by 5)
- As a rate, such as 0.4 (i.e., the result of the division expressed by the fraction above)
- As a percentage, such as 40% (i.e., the rate above multiplied by 100, followed by a percent sign)

Each of these expressions is useful in different contexts, but rates and percentages are the most concise and therefore the most useful for tables and graphs. Many measures of ratio have conventional forms of expression, such as the book-to-bill rate mentioned above, which is typically expressed as a rate (e.g.,

1.25, which indicates that for every five orders that have booked, only four have been billed, or 5 ÷ 4 = 1.25), or the profit margin, which is normally expressed as a percentage (e.g., 25%, which indicates that for every $100 of revenue, $75 goes toward expenses, leaving a profit of $25, or $25 ÷ $100 = 0.25 × 100 = 25%).

Take a moment to think about and list a few of the ways that ratios are used, or could be used, to present quantitative information related to your own work.

. . . .

Ratios are simple shorthand for expressing the direct relationship between two quantitative values. One especially handy use of ratios is to compare several individual values to a particular value to show how they differ. In this case, your purpose is not to compare the actual values but to show the degree to which they differ. In such circumstances, you can simplify the message by setting the main value to which you are comparing all the others to 1 (expressed as a rate) or 100% (expressed as a percentage); then, express the other values as ratios that fall above or below that point of reference. Here's an example expressed in percentages:

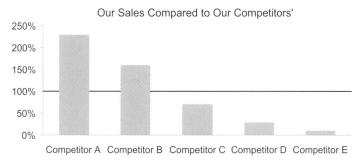

FIGURE 2.21 This graph includes a reference point of 100% for the primary set of values, making it easy to see how the other values, also expressed as percentages, differ.

Using 100% as a consistent point of reference, it is easy to see that the main competitor's sales are about 230% of your company's, or 2.3 times greater when expressed as a rate. Expressing comparisons in this manner eliminates the need for readers to do calculations in their heads when they care about relative differences.

Measures of Money

Much of the quantitative information that's important in business involves some currency of exchange—in other words, money. Be it U.S., Canadian, or Australian dollars, Japanese yen, British pounds, Swiss francs, or Euros, money is at the center of most business analysis and reporting. Unlike most other units of measure, currency has a characteristic that we must keep in mind when communicating information that spans time: the value of money is not static; it changes with time. The value of a U.S. dollar in November of 2001 was not the same as its value in November of 2010. If you've been asked to prepare a report that exhibits the trend of sales in U.S. dollars for the past 10 years, would you be justified in asserting that sales have increased by 100% during that time if 10 years ago annual sales were $100 million and today they total $200 million? That assertion would be true only if the value of a dollar today is the same as it was 10 years ago, which it isn't.

When the value of a dollar decreases over time, we refer to this as *inflation*. We can accurately compare money over time only if we adjust for inflation. I've noticed, however, that this is rarely done. Despite the validity of the case for adjusting for inflation, doing so isn't always practical, so I won't try to force on you a practice that you might very well ignore. For those of you who can take extra time required to correct for results skewed by inflation, I've included Appendix C, *Adjusting for Inflation*, in the back of the book. Adjusting for inflation isn't difficult, and doing so will improve the quality of your financial reporting.

Business today, especially in large companies, is often international and involves multiple currencies. This is a problem when we must produce reports that combine data in multiple currencies, such as sales in the Americas, Europe, and Asia. You can't just throw the numbers together because 100,000 U.S. dollars does not equal 100,000 British pounds or 100,000 Japanese yen. To combine or compare them, you must convert them into a single currency. Fortunately, most software systems that we use today are designed to do this work for us, converting money based on tables of exchange rates, so we can easily see transactions both in their original currency and in some common currency used for international reporting, such as U.S. dollars. Because software typically does this work for us, my intention here is simply to caution you to avoid mixing currencies without converting them to a common currency. If you're not careful, you could inadvertently report results that are in error by a large order of magnitude.

Understanding the relationships we've examined in this chapter lays a foundation that will help you design tables and graphs to effectively communicate quantitative information. In the next chapter, we'll look at the basics of tables and graphs and begin to see how they can effectively present the kinds of relationships we've just discussed.

Summary at a Glance

Quantitative Relationships

- Quantitative stories include two types of values:
  - Quantitative
  - Categorical
- Quantitative stories always feature relationships.
- These relationships involve either:
  - Simple associations between quantitative values and categorical items or
  - More complex associations among multiple sets of quantitative values.

- There are four types of relationships between categorical items:
  - Nominal
  - Ordinal
  - Interval
  - Hierarchical
- There are three types of relationships between quantitative values:
  - Ranking
  - Ratio
  - Correlation

Numbers that Summarize

Type of Summary	Method	Note
Average	Mean	Measures the center of a set of values in a manner that is equally sensitive to all values, including extremes
	Median	Measures the center of a set of values in a manner that is insensitive to extreme values
Variation	Spread	Simple to calculate, relying entirely on the highest and lowest values, but only roughly defines a range of values
	Standard Deviation	Provides a rich expression of the distribution of a set of values across its entire range
Correlation	Linear Correlation Coefficient	Indicates whether a linear correlation exists between two paired sets of values, and if so, its direction (positive or negative) and its strength (strong or weak)
Ratio	Rate or Percentage	Measures the direct relationship between two quantitative values

Measures of Money

- When comparisons of monetary value are expressed across time, adjusting the value to account for inflation produces the most accurate results.
- When reporting monetary values that combine multiple currencies, you must first convert them all into a common currency.

3 DIFFERING ROLES OF TABLES AND GRAPHS

Tables and graphs are the two fundamental vehicles for presenting quantitative information. They have developed over time to the point that we now thoroughly understand which works best for different circumstances and why. This chapter introduces tables and graphs and gives simple guidelines for selecting which to use for your particular purpose.

Tables and graphs are the two primary means to structure and communicate quantitative information. Both have been around for quite some time and have been researched extensively to hone their use to a fine edge of effectiveness. The best practices that have emerged are easy to learn, understand, and put to use in your everyday work with numbers.

Occasionally, the best way to display quantitative information is not in the form of a table or graph. When the quantitative information you want to convey consists only of a single number or two, written language is an effective means of communication; your message can be expressed simply as a sentence or highlighted as a bullet point. If your message is that last quarter's sales totaled $1,485,393 and exceeded the forecast by 16%, then it isn't necessary to structure the message as a table, and there is certainly no need to create a graph. You can simply say something like:

Q2 sales = $1,485,393, exceeding forecast by 16%

Alternatively, it wouldn't hurt to structure this message in simple tabular form such as this:

Q2 Sales	Compared to Forecast
$1,485,393	+16%

FIGURE 3.1 This table shows sales value information, arranged in columns.

or this:

Q2 Sales	$1,485,393
Compared to Forecast	+16%

FIGURE 3.2 This table shows sales value information, arranged in rows.

If you're like a lot of folks, however, you might be tempted while structuring this information as a table to jazz it up in a way that actually distracts from your simple, clear message—perhaps something like this:

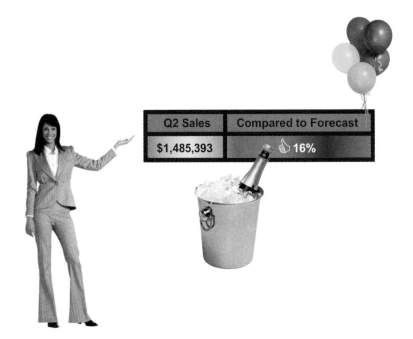

FIGURE 3.3 This table shows sales values, designed to impress, or perhaps to entertain, but not primarily to communicate.

You might even be tempted to pad the report with an inch-thick stack of pages containing the details of every sales order received during the quarter, eager to demonstrate how hard you worked to produce those two sales numbers. However, as we'll observe many times in this book, there is *eloquence in simplicity.*

Quantities and Categories

Before we launch into an individual examination of tables and graphs, let's review an important fact, common to both, that quantitative messages are made up of two types of data:

- Quantitative
- Categorical

Quantitative values measure something (number of orders, amount of profit, rating of customer satisfaction, etc.). Categorical items (i.e., members of a category) identify what the quantitative values measure. These two types of data fulfill different roles in tables and graphs. In the following simple table, which displays exempt and non-exempt employee compensation by department, all the information consists either of quantities being measured (compensation) or items belonging to categories (sales, operations, and manufacturing are items in the department category) to which the quantities relate.

Department	Exempt	Non-Exempt
Sales	950,003	1,309,846
Operations	648,763	2,039,927
Manufacturing	568,543	2,367,303
Total	$2,167,309	$5,717,076

FIGURE 3.4 This is a table of employee compensation information that you can use to practice distinguishing quantitative and categorical data.

The labels that identify the various departments in the far left column, including Total, belong to the category Department, and the labels Exempt and Non-exempt belong to a separate category that we could call Employee Type. All of the other data (i.e., all the numbers in this table) are quantitative values.

Do numbers always represent quantitative values? No. Sometimes numbers are used to label things and have no quantitative meaning. Order numbers (e.g., 1003789), numbers that identify the year (e.g., 2011), and numbers that sequence information (item number 1, item number 2, item number 3, etc.), are examples of numbers that express categorical data. One simple test to determine this distinction is to ask the question, "Would it make sense to add these numbers up, or to perform any mathematical operation on them?" For instance, would it make sense to add up order numbers? No, it wouldn't. How about numbers that rank a list of sales people as number 1, number 2, etc.? Once again, the answer is no; therefore, these numbers represent categorical, rather than quantitative, data.

Let's take a look at one more example to practice making the distinction between categorical and quantitative information. Using the graph below, test your skills by identifying which components represent categorical items and which represent quantitative values.

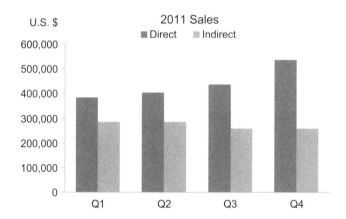

FIGURE 3.5 This graph depicts sales information for practice in distinguishing quantitative and categorical data.

Let's examine one component at a time. For each of the items below, indicate whether the information is categorical or quantitative.

1. The values of time along the bottom (Q1, Q2, etc.)
2. The dollars along the left side
3. The legend, which encodes Direct and Indirect
4. The vertical bars in the body of the graph
5. The title in the top center

Once you've identified each, take a moment to compare your answers to those on the right.

. . . .

Did you catch the dual role of the bars, that they contain both quantitative values and categorical items? With a little practice, you will be able to easily

ANSWERS
1. Categorical, labeling the quarters of the year
2. Quantitative, providing dollar values for interpreting the heights of the bars
3. Categorical, providing a distinction between direct and indirect sales
4. Both quantitative and categorical; the heights of the bars encode quantitative information about sales in dollars; the colors of the bars encode categorical data identifying which sales are direct vs. indirect
5. Categorical, identifying the year of the sales

deconstruct graphs into quantitative and categorical data. This ability will enable you to apply the differing design practices that pertain to each type of data.

Choosing the Best Medium of Communication

Choosing whether to display data in one or more tables, one or more graphs, or some combination of the two, is a fundamental challenge of data presentation. This decision should never be arbitrary. It is sad, however, that this choice is often made using the "eeney, meeney, miney, moe" method. Imagine that you're Joe, and you've just interviewed three candidates for a new position in your department. It's now time to report the results to your boss, who's responsible for making the hiring decision. During interviews with the candidates, you used your company's handy interview score sheet to evaluate each person's aptitudes on a zero to five point scale in six areas: experience, communication, friendliness, subject matter knowledge, presentation, and education. How do you display the results? Well, if you're like a lot of people who feel pressure to make an impression, you might decide to use something unusual, like one of those radar charts that are available in your spreadsheet software. As you hand your boss the following report, you hope that he's thinking: "Wow, that Joe is an exceptional employee. Perhaps he deserves a raise."

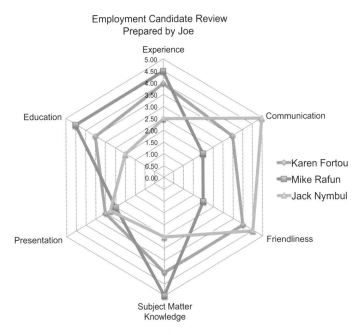

FIGURE 3.6 This radar chart presents aptitudes of job candidates in an overly complicated way.

What's sad is that the boss, upon first glance, might actually think, "That Joe certainly outdid himself," despite the fact that he hasn't a clue how to read this spider-web of a chart. He might actually blame himself for lacking the skill that's needed to read this chart, assuming he must have missed that day in math class when they covered this. Regardless of how the boss responds, Joe would have made his job easier if he'd prepared something like this instead:

Employment Candidate Review

Rating Areas	Candidates		
	Karen Fortou	Mike Rafun	Jack Nymbul
Experience	4.00	4.50	2.50
Communication	3.50	2.00	5.00
Friendliness	4.00	2.00	4.50
Subject matter knowledge	4.00	5.00	2.50
Presentation	3.00	1.50	2.75
Education	3.50	4.50	2.00
Average Rating	3.67	3.25	3.21

FIGURE 3.7 This table presents the same information as Figure 3.6 but in a way that is simple and clear.

This table presents the information in a way that is simple to understand and efficient to use. To select the appropriate medium of communication, we must understand the needs of our audience as well as the purposes for which various forms of display can be effectively used. Let's begin by considering the overall strengths and weaknesses of tables and graphs.

Tables Defined

A table is a structure for organizing and displaying information; a table exhibits the following characteristics:

- Information is arranged in columns and rows.
- Information is encoded as text (including words and numbers).

A single set of values, occupying a single column or row, is merely a list, not a table.

Although the column and row structure of tables is often visually reinforced by grid lines (i.e., horizontal and vertical lines outlining the columns and rows), it is the arrangement of the information that characterizes tables, not the presence of lines that visibly delineate the structure of the underlying grid. In fact, as we will see later in the chapter on table design, grid lines must be used with care to keep them from diminishing a table's usefulness.

Tables are not used exclusively to display quantitative information. Whenever you have more than one set of values, and a relationship exists between values in the separate sets, you may use a table to align the related values by placing them in the same row or column. For instance, tables are often used to display meeting agendas, with start times in one column, the names of the topics that will be covered in the next, and the names of the facilitators in the next, as in the following example:

Time	Topic	Facilitator
09:00 AM	Opening remarks	Scott Wiley
09:15 AM	Product demo	Sheila Prescott
10:00 AM	Discussion	Jerry Snyder
10:45 AM	Planning	Pamela Smart

FIGURE 3.8 This is an example of a table that does not contain quantitative data, in this case a meeting agenda.

Tables have been in use for almost two millennia, so they are readily understood by almost everyone who can read.

When to Use Tables

A handful of conditions should direct you to select a table, rather than a graph, as the appropriate means of display, but tables provide one primary benefit:

Tables make it easy to look up individual values.

Tables excel as way to display simple relationships between quantitative values and the categorical items to which they're related so that these individual values can be easily located.

When deciding whether to use a table or graph to communicate your quantitative message, always ask yourself how the information will be used. Will you or others use it to look up one or more particular values? If so, it's a prime candidate for expression in a table. Or might the information be used to examine a set of quantitative values as a whole to discern patterns? If so, it's a prime candidate for expression in a graph, as we'll see soon.

Tables also make it easy to compare pairs of related values (e.g., sales in quarter 1 to sales in quarter 2). Here's a typical example:

Year	Jan	Feb	Mar	Apr	May	Jun	Jul	Aug	Sep	Oct	Nov	Dec	Monthly Average
2000	138.1	138.6	139.3	139.5	139.7	140.2	140.5	140.9	141.3	141.8	142.0	141.9	**140.3**
2001	142.6	143.1	143.6	144.0	144.2	144.4	144.4	144.8	145.1	145.7	145.8	145.8	**144.5**
2002	146.2	146.7	147.2	147.4	147.5	148.0	148.4	149.0	149.4	149.5	149.7	149.7	**148.2**
2003	150.3	150.9	151.4	151.9	152.2	152.5	152.5	152.9	153.2	153.7	153.6	153.5	**152.4**
2004	154.4	154.9	155.7	156.3	156.6	156.7	157.0	157.3	157.8	158.3	158.6	158.6	**156.9**
2005	159.1	159.6	160.0	160.2	160.1	160.3	160.5	160.8	161.2	161.6	161.5	161.3	**160.5**
2006	161.6	161.9	162.2	162.5	162.8	163.0	163.2	163.4	163.6	164.0	164.0	163.9	**163.0**
2007	164.3	164.5	165.0	166.2	166.2	166.2	166.7	167.1	167.9	168.2	168.3	168.3	**166.6**
2008	168.8	169.8	171.2	171.3	171.5	172.4	172.8	172.8	173.7	174.0	174.1	174.0	**172.2**
2009	175.1	175.8	176.2	176.9	177.7	178.0	177.5	177.5	178.3	177.7	177.4	176.7	**177.1**
2010	177.1	177.8	178.8	179.8	179.8	179.9	180.1	180.7	181.0	181.3	181.3	180.9	**179.9**

FIGURE 3.9 This is an example of a simple table that can be used to look up several years of monthly rates.

Tables work well for look-up and one-to-one comparisons, in part because their structure is so simple, and in part because the quantitative values are encoded as text, which we can understand directly, without translation. Graphs, by contrast, are visually encoded, which requires translation of the information into the numbers it represents.

The textual encoding of tables also offers a level of precision that cannot be provided by graphs. It is easy to express a number with as much specificity as you wish using text (e.g., 27.387483), but the visual encoding of individual numbers in graphs doesn't lend itself to such precision.

Another strength of tables is that they can include multiple sets of quantitative values that are expressed in different units of measure. For instance, if you need to provide sales information that includes the number of units sold, the dollar amount, and a comparison to a forecast expressed as a percentage, doing so in a single graph would be difficult, because a graph usually contains a single quantitative scale with a single unit of measure.

And a final strength of tables is their ability to combine summary and detail information in a single display. For example, a table might include the amount of revenue earned per month (detail), with a total for the year (summary).

To summarize, tables are used to display simple relationships between quantitative values and corresponding categorical items, which makes tables ideal for looking up and comparing individual values. The entire set of reasons to use a table consists of the following; if one or more of these are true, you should probably display the data in a table:

1. The display will be used to look up individual values.
2. It will be used to compare individual values but not entire series of values to one another.
3. Precise values are required.
4. The quantitative information to be communicated involves more than one unit of measure.
5. Both summary and detail values are included.

Graphs Defined

Graphs exhibit the following characteristics:

- Values are displayed within an area delineated by one or more axes.
- Values are encoded as visual objects positioned in relation to the axes.
- Axes provide scales (quantitative and categorical) that are used to label and assign values to the visual objects.

Axes delineate the space that is used to display data in a graph.

Essentially, a graph is a *visual display of quantitative information*. Whereas tables encode quantitative values as text, graphs encode quantitative values visually. Consider this simple example:

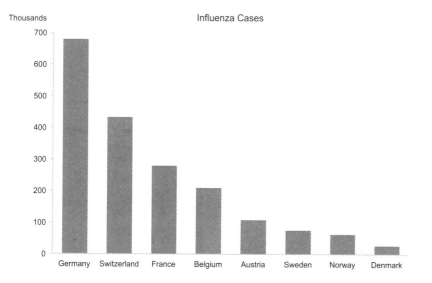

FIGURE 3.10 This is an example of a simple graph, which displays a count of influenza cases per country.

This graph has two axes: one that runs horizontally, called the *X axis*, and one that runs vertically, called the *Y axis*. In this graph, the categorical scale, which labels the countries, resides along the X axis, and the quantitative scale (i.e., counts of influenza cases) resides along the Y axis. The values themselves are encoded as rectangles, called *bars*. Bars are one of several visual objects that can be used to encode data in graphs. The number of influenza cases in each country is encoded as the height of its bar and the position of its top in relation to the scale on the Y axis. The horizontal position of each bar along the X axis is labeled to denote the specific categorical item to which the values are related (e.g., Belgium).

With a little practice, even someone who has never previously used a graph can learn to interpret the information contained in a simple one like this. Although the values of sales for each country cannot be interpreted to the exact number, this isn't the graph's purpose. Rather, the graph paints the picture that influenza has affected these countries to varying degrees, with the greatest effect in Germany and the least in Denmark, and significant differences in between. Information like this, which is intended to show patterns in data, is best presented in a graph rather than a table, as we'll see in the section *When to Use Graphs*. But first, a little history.

A Brief History of Graphs

Graphs of quantitative information have been in use for only a few hundred years, which is a relatively short time given the thousands of years that mathematics has existed. Despite how natural it may seem to see quantitative information displayed in graphs, the original notion that numbers could be displayed visually in relation to two perpendicular axes involved a leap of imagination. The launching pad for this leap had already been around for many centuries before quantitative graphs emerged. An earlier type of visual information display, also assisted by a scale of measurement along perpendicular axes, eventually suggested the possibility of graphs. Can you guess what it was? It is still in common use today for the purpose of navigation. It is a two-dimensional representation of the physical world that's used to measure distances between locations. I'm referring to a *map*. The earliest known map dates back about 4,300 years. It was drawn on a clay tablet and represented northern Mesopotamia. When a map depicts the entire world, or some large part of it, the standard set of grid lines that allow us to determine location and distance are longitudes and latitudes. These grid lines form a quantitative scale of sorts.

In the 17th century these grids were adapted for the representation of numbers alone. In his *La Géométrie* (1637), René Descartes introduced them as a means to encode numbers as coordinate positions in a two-dimensional (2-D) grid. This innovation provided the groundwork for an entire new field of mathematics that is based on graphs. To some extent even prior to Descartes, others had already experimented with graphical displays of numbers. Ron Rensink writes:

> Although Descartes did contribute to the graphic display of quantitative data in the 17th century, other forms of quantitative graphs had already been used to represent things such as temperature and light intensity three centuries earlier. Indeed, as Manfredo Massironi discusses in his book (Massironi, M. (2002) The Psychology of Graphic Images: Seeing, Drawing, Communicating. Erlbaum, page 131), quantities such as displacement were graphed as a function of time as far back as the 11th century. But while these facts may be of interest in their own right, the more important point

*is that techniques in graphic representation have been developed over many
centuries, and many of these techniques have been subsequently forgotten—perhaps fallen out of vogue, or never found wide use to begin with.
But the reasons for their dismissal may not necessary apply in this day and
age. Indeed, several techniques might lend themselves quite well to modern
technology, and so might be worth resurrecting in one form or other. Books
such as Massironi's are helpful in discovering such possibilities.* [1]

Despite these earlier ventures, it wasn't until the late 18th century that the use
of graphs to present numbers became popular. William Playfair, a Scottish
social scientist, used his imagination and design acumen to invent many of the
graphing techniques that we use today. He pioneered the use of graphs to reveal
the shape of quantitative information, thus providing a way to communicate
quantitative relationships that numbers expressed as text could never convey. He
invented the bar graph, was perhaps the first person to use a line to show how
values change through time, and on a day when he was probably under the
weather, invented the pie chart as well, but we'll forgive his brief lapse of
judgment.

1. Ron Rensink (2011)
"Four Futures and a History,"
www.Interaction-Design.org.

Information about the early
development of the graph, including
its precursor, the map, may be found
in Robert E. Horn (1998) *Visual
Language.* MacroVU, Inc. Robert
Horn provides an informative
timeline, which cites the milestones
in the historical development of
visual information display.

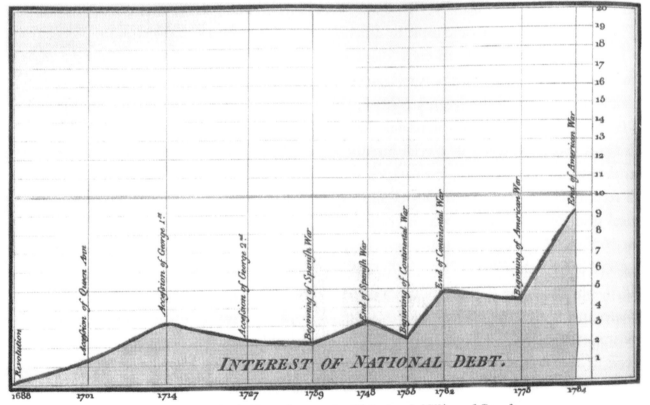

The old saying, "A picture is worth a thousand words" applies quite literally to
graphs. By presenting quantitative information in visual form, graphs efficiently
communicate what might otherwise require a thousand or even a million words,
and sometimes communicate what words could never convey.

FIGURE 3.11 This graph was
included in William Playfair's book
The Commercial and Political Atlas in
1786 to make a case against
England's policy of financing colonial
wars through national debt.

From the time of Playfair until today, many innovators have added to the inventory of graph designs available for the representation, exploration, analysis, and communication of quantitative information. During the past 50 years, none has contributed more to the field as an advocate of excellence in graphic design than Edward Tufte, who in 1983 published his landmark treatise on the subject, *The Visual Display of Quantitative Information*. With the publication of additional books and articles since, Tufte continues to be respected as a major authority in this field.

The work of William S. Cleveland, especially his book *The Elements of Graphing Data*, is also an outstanding resource.[2] Cleveland's work is particularly useful to those with statistical training who are interested in sophisticated graphs, such as those used in scientific research.

2. William S. Cleveland (1994) *The Elements of Graphing Data*. Hobart Press.

When to Use Graphs

Graphs reveal more than a collection of individual values. Because of their visual nature, graphs present the overall shape of the data. Text, displayed in tables, cannot do this. The patterns revealed by graphs enable readers to detect many points of interest in a single collection of information. Take a look at the next example, and try to identify some of the features that graphs can reveal. Approach this by first determining the messages in the data, then by identifying the visual cues—the shapes—that reveal each of these messages.

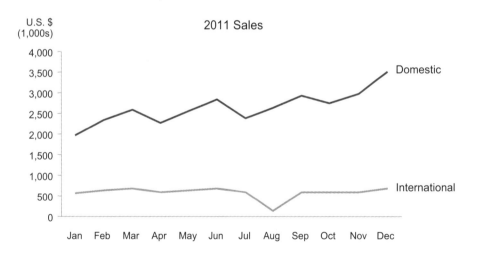

FIGURE 3.12 This fairly typical line graph is an example of the shapes and patterns in quantitative data that graphs make visible.

What insights are brought to your attention by the shape of this information? Take a moment to list them in the right margin.

· · · · · · · ·

Let's walk through a few of these revelations together, beginning with domestic sales. What does the shape of the black line tell you about domestic sales during the year 2011? One message is that, during the year as a whole, domestic sales increased; these sales ended higher than they started, with a gradual increase through most of the year. One name for this type of pattern is a *trend*, which displays the overall nature of change during a particular period of time.

Was this upward trend steady? No, it exhibits a pattern that salespeople sometimes call the *hockey stick*. Sales go down in the first month of each quarter and then gradually go up to a peak in the last month of each quarter. If you examine the shape of the line from the last month of one quarter, such as March, to the last month of the next quarter, such as June, you'll recognize that these segments each look a little like a hockey stick. This pattern in sales usually occurs when salespeople are given bonuses based on reaching or exceeding a quarterly quota.

Now, if you look at sales in the first quarter of the year versus sales in the second quarter, then the third, and finally the fourth, you find that the graph makes it easy to see how these different periods compare. When information is given shape in a graph, it becomes possible to compare entire sets of data. In this case the comparison is between different quarters.

Now look for a moment at international sales. Once again you are able to easily see the trend of sales throughout the year, which in this case is relatively flat. Compared with domestic sales, international sales seem to exhibit less fluctuation through time and relatively little difference between the beginning and end of the year. Although international sales appear to be less affected by quarters and seasons, one point along the line stands out as quite different from the rest: the month of August. Sales in August took an uncharacteristic dip compared to the rest of the year. As an analyst, this abnormal sales value would make you want to dig for the cause. If you did, perhaps you would discover that most of your international customers were vacationing in August and therefore weren't around to place orders. Whatever the cause, the current point of interest to us is that graphs make exceptions to general patterns stand out clearly from the rest. This wouldn't be nearly so obvious in a table of numbers.

Finally, if you widen your perspective to all the quantitative information, you find that the graph makes it easy to see the similarities and differences between the two sets of values (domestic and international sales), both overall and at particular points in the graph. We could go on, adding more to the list of characteristics that graphs reveal, but we've hit the high points and are now ready to distill what we've detected to its essence:

> Graphs are used to display relationships among and between
> sets of quantitative values by giving them shape.

The visual nature of graphs endows them with their unique power to reveal patterns of various types, including changes, differences, similarities, and exceptions. Graphs can communicate quantitative relationships that are much more complex than the simple associations between individual quantitative values that tables can express.

Graphs can display large data sets in a way that can be readily perceived and understood. You could gather data regarding the relationship between employee productivity and the use of two competing software packages, involving thousands of records across several years, and, with the help of a graph, you would be able to immediately see the nature of the relationship. If you have ever tried to use a huge table of data for analysis, you would quickly fall in love with graphs like scatter plots, which can make the relationships among thousands of individual data points instantly intelligible.

Patterns in data are difficult to discern even in a small table of numbers. Despite our agreement on most matters, Tufte and I disagree slightly about this. According to Tufte:

> *Tables usually outperform graphics in reporting on small sets of 20 numbers or less. The special power of graphics comes in the display of large data sets.* [3]

3. Edward Tufte (2001) *The Visual Display of Quantitative Information*, Second Edition. Graphics Press, page 56.

Although it is true that one's chance of seeing a meaningful pattern in small table of numbers is greater than in one that is large, those patterns are much easier to see in a graph. Take a look at the following table.

Job Satisfaction By Income, Education, and Age

	College Degrees		No College Degrees	
Income	Under 50	50 & over	Under 50	50 & over
Up to $50,000	643	793	590	724
Over $50,000	735	928	863	662

FIGURE 3.13 This table displays measures of employee job satisfaction, divided into categories.

Do any unusual patterns of job satisfaction pop out? Can you see that one group of employees exhibits a different pattern of satisfaction from the others? Now take a look at the following graph.

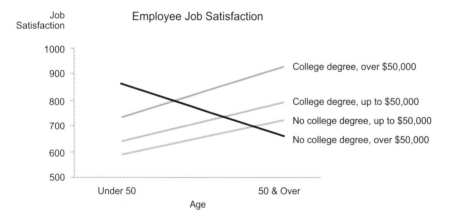

FIGURE 3.14 This graph causes the pattern represented by the black line to pop out.

The fact that employees with salaries over $50,000 but no college degrees experience a significant decrease in job satisfaction in their older years now jumps out. Even with only eight numbers, this graph did what the table couldn't do—it made this diverging pattern obvious. This results not only because the information is displayed graphically, but also because the graph was specifically designed to feature this particular pattern. Had I designed the graph in the following manner, that pattern would no longer be obvious.

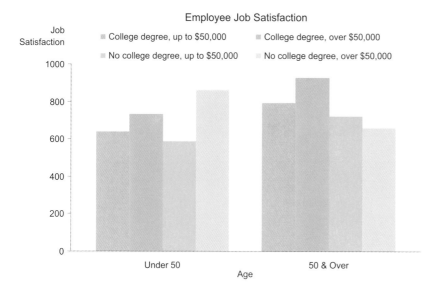

FIGURE 3.15 Unlike the graph in *Figure 3.14*, this graph does not make the pattern of decreasing job satisfaction among the one group of employees obvious.

This example is not based on real data, so if you make more than $50,000 a year but don't have a college degree and will soon turn 50, don't worry, you'll probably do just fine.

So displaying information in a graph rather than a table does not by itself make meaningful patterns visible. We must design the graph to feature evidence of the particular story that we're trying to tell. Many possible stories dwell in a set of data, even in a small one such as illustrated above. Later in this book, we'll learn to make the necessary graph design choices to tell particular stories.

Summary at a Glance

Use Tables When	Use Graphs When
• The display will be used to look up individual values. • It will be used to compare individual values. • Precise values are required. • The quantitative values include more than one unit of measure. • Both detail and summary values are included.	• The message is contained in the shape of the values (e.g., patterns, trends, and exceptions). • The display will be used to reveal relationships among whole sets of values.

4 FUNDAMENTAL VARIATIONS OF TABLES

Tables should be structured to suit the nature of the information they are meant to display. This chapter breaks tables down into their fundamental variations and provides simple rules of thumb for pairing your message with the best tabular means to communicate it.

We'll begin this chapter by identifying *what* tables can be used to display, followed by *how* they can be structured visually, and then we'll link the *what* to the *how*.

Relationships in Tables

Information that we display in tables always exhibits a specific relationship between individual values. Tables are commonly used to display one of five potential relationships, which can be divided into two major types:

- Quantitative-to-categorical relationships
  - Between one set of quantitative values and one set of categorical items
  - Between one set of quantitative values and the intersection of multiple categories
  - Between one set of quantitative values and the intersection of hierarchical categories
- Quantitative-to-quantitative relationships
  - Among one set of quantitative values associated with multiple categorical items
  - Among distinct sets of quantitative values associated with a single categorical item

This might sound like gobbledygook, but the meaning of each relationship will become clear as we proceed.

Quantitative-to-Categorical Relationships

Tables of this type are primarily used for looking up one quantitative value at a time; each value relates either to a single category or to the intersection of multiple categories.

BETWEEN ONE SET OF QUANTITATIVE VALUES AND ONE SET OF CATEGORICAL ITEMS
The simplest type of relationship displayed in tables is an association between one set of quantitative values and one set of categorical items. Typical examples include sales dollars (quantitative values) associated with geographical sales regions (categorical items), ratings of performance (quantitative values)

associated with individual employees (categorical items), or expenses (quantita-tive values) associated with fiscal periods (categorical items). Here's an example:

Salesperson	QTD Sales
Robert Jones	13,803
Mandy Rodriguez	20,374
Terri Moore	28,520
John Donnelly	34,786
Jennifer Taylor	36,973
Total	$134,456

FIGURE 4.1 This table displays a simple relationship between quantitative values and categorical items, in this case sales in dollars per salesperson.

Tables of this type function as simple lists used for look-up. You want to know how much Mandy has sold so far this quarter? You simply find Mandy's name on the list, and then look at the sales number next to her name. Tables of this sort are a common means of breaking quantitative information down into small chunks so we can see the details behind the bigger picture.

BETWEEN ONE SET OF QUANTITATIVE VALUES AND THE INTERSECTION OF MULTIPLE CATEGORIES

One small step up the complexity scale, tables can also display quantitative values that are simultaneously associated with multiple sets of categorical items. Here's the same example as the one above, but this time a second category has been added:

Salesperson	Jan	Feb	Mar
Robert Jones	2,834	4,838	6,131
Mandy Rodriguez	5,890	6,482	8,002
Terri Moore	7,398	9,374	11,748
John Donnelly	9,375	12,387	13,024
Jennifer Taylor	10,393	12,383	14,197
Total	$35,890	$45,464	$53,102

FIGURE 4.2 This table displays a relationship between quantitative values and multiple categories, in this case between sales dollars and particular salespeople (category 1) in a particular month (category 2).

In this example, each quantitative value measures the sales for a particular salesperson in a particular month. A single quantitative value measures sales for the intersection of two categories (salesperson and month). This is still a rela-tionship between quantitative values and categorical items, but it is slightly more complex in that it addresses the intersection of multiple categories. Because each category represents a different dimension of the information, this type of table provides a multi-dimensional view.

BETWEEN ONE SET OF QUANTITATIVE VALUES AND THE INTERSECTION OF HIERARCHICAL CATEGORIES

A variation of the relationship shown in the previous example is when multiple sets of categorical items relate to one another hierarchically. For instance, let's consider the following three categories of time: years, quarters, and months. They are related hierarchically in that years consist of quarters, which, in turn, consist of months. Another common example involves the hierarchical categories that organize and describe products. These might consist of product lines, which consist of product families, which consist of individual products. Just as in the previous type of relationship, a quantitative value measures the intersection of multiple categories. The only difference is that in this case there is a hierarchical relationship among the categories. Here's an example:

Product Line	Product Family	Product	Sales
Hardware	Printer	PPS	6,131
		PXT	8,002
		PQT	11,748
	Router	RRZ	13,024
		RTS	14,197
		RQZ	23,293
Software	Business	ACT	12,393
		SPR	9,393
		DBM	5,392
	Game	ZAP	10,363
		ZAM	15,709
		ZOW	13,881
Total			$143,526

FIGURE 4.3 This table displays a relationship between quantitative values and hierarchical categories, in this case between sales dollars and a hierarchically arranged combination of product line, product family, and product.

Quantitative-to-Quantitative Relationships

In addition to permitting us to show direct relationships between quantitative values and categorical items, tables also allow us to examine relationships among multiple quantitative values. These relationships are primarily used to compare multiple quantitative values of a single measure (e.g., expenses in U.S. dollars) associated with multiple categorical items (e.g., the months January and February), or multiple measures (e.g., expenses and revenues in U.S. dollars) for a single categorical item (e.g., the month of January).

AMONG ONE SET OF QUANTITATIVE VALUES ASSOCIATED WITH MULTIPLE CATEGORICAL ITEMS

Rather than merely looking at the sales associated with an individual salesperson, you might want to compare sales among several salespeople. In the next example, how did Mandy do compared to Terri in February?

Salesperson	Jan	Feb	Mar
Robert Jones	2,834	4,838	6,131
Mandy Rodriguez	5,890	6,482	8,002
Terri Moore	7,398	9,374	11,748
John Donnelly	9,375	12,387	13,024
Jennifer Taylor	10,393	12,383	14,197
Total	$35,890	$45,464	$53,102

FIGURE 4.4 This table examines a relationship among quantitative values associated with multiple categorical items, in this case sales by several salespeople in the months of January, February, and March.

As you can see, this table is identical in design to the table in *Figure 4.2*, but in this case the primary activity is comparison rather than look-up.

AMONG DISTINCT SETS OF QUANTITATIVE VALUES ASSOCIATED WITH A SINGLE CATEGORICAL ITEM

Tables often contain more than one set of quantitative values (i.e., distinct measures). In this case, we can examine relationships among different quantitative measures that are associated with a single categorical item. The table below extends the previous example to three distinct measures: sales, returns, and net sales:

Salesperson	Sales	Returns	Net Sales
Robert Jones	13,803	593	13,210
Mandy Rodriguez	20,374	1,203	19,171
Terri Moore	28,520	10,393	18,127
John Donnelly	34,786	483	34,303
Jennifer Taylor	36,973	0	36,973
Total	$134,456	$12,672	$121,784

FIGURE 4.5 This table examines a relationship among distinct measures associated with a single categorical item, in this case the salespeople's sales, returns, and resulting net sales.

This table makes it easy to see how each salesperson is doing by comparing three measures of performance.

Here's a summary of *what* tables can be used to display:

Primary Function	Relationship Type	Relationship
Look-up	Quantitative-to-Categorical	Between a single set of quantitative values and a single set of categorical items
		Between a single set of quantitative values and the intersection of multiple categories
		Between a single set of quantitative values and the intersection of multiple hierarchical categories
Comparison	Quantitative-to-Quantitative	Among a single set of quantitative values associated with multiple categorical items
		Among distinct sets of quantitative values associated with a single categorical item

It is important to distinguish these relationships because, as we'll see in the next section, the layout of a table is primarily determined by the relationships the table is meant to feature.

Variations in Table Design

Now let's look at *how* tables can display the types of relationships we've identified. Within the general structure of columns and rows, tables can vary somewhat in design. Structural variations can be grouped into two fundamental types:

- Unidirectional—categorical items are laid out in one direction only (i.e., either across columns or down rows)
- Bidirectional—categorical items are laid out in both directions

Unidirectional

When a table is structured unidirectionally, categorical items are arranged across the columns or down the rows but not in both directions. Unidirectional tables may have one or more sets of categorical items, but if there is more than one, the values are still only arranged across the columns or down the rows. Here's an example of a unidirectional table with categorical items (i.e., departments) arranged down the rows:

Department	Headcount	Expenses
Finance	26	202,202
Sales	93	983,393
Operations	107	933,393
Total	234	$2,118,988

FIGURE 4.6 This table is structured unidirectionally with the categorical items (departments) arranged down the rows.

Notice that Headcount and Expenses label two distinct sets of quantitative values, not values of a single category. Here's the same information, this time with the categorical items (i.e., departments) laid out across the columns:

Dept	Finance	Sales	Ops	Total
Headcount	26	93	107	234
Expenses	202,202	983,393	933,393	$2,118,988

FIGURE 4.7 This table is structured unidirectionally with the categorical items arranged across the columns.

These are simple unidirectional tables in that they only contain a single set of categorical items, but tables can also have multiple sets, as in the following example:

Dept	Expense Type	Expenses
Finance	Compensation	160,383
	Supplies	5,038
	Travel	10,385
Sales	Compensation	683,879
	Supplies	193,378
	Travel	125,705
Total		$1,178,768

FIGURE 4.8 This table is structured unidirectionally, consisting of two sets of categorical items: departments in the left column and expense types in the middle column.

Because the two sets of categorical items are arranged in only one direction, in this case down the rows, this table is still structured unidirectionally.

Bidirectional

When a table is structured bidirectionally, more than one set of categorical items is displayed, and these sets are laid out both across the columns and down the rows. This arrangement is sometimes called a *crosstab* or a *pivot table*. The quantitative values appear in the body of the table, bordered by the categorical items, which run across the top and along the left side. This type of table can be easily understood by looking at an example. Here's the same information found in the previous unidirectional table, but this time it is structured bidirectionally:

	Departments		
Expense Types	Finance	Sales	Total
Compensation	160,383	683,879	844,262
Supplies	5,038	193,378	198,416
Travel	10,385	125,705	136,090
Total	$175,806	$1,002,962	$1,178,768

FIGURE 4.9 This is an example of a bidirectionally structured table.

To locate the quantitative value associated with a particular department and a particular expense type, you look where the relevant column and relevant row intersect.

One of the advantages of bidirectional tables over unidirectional tables is that they can display the same information using less space. Here are the same examples of the previous unidirectional and bidirectional tables, but this time I've included a complete set of grid lines to make it easy to see the amount of space used by each. Here's the unidirectional version:

Dept	Expense Type	Expenses
Finance	Compensation	160,383
	Supplies	5,038
	Travel	10,385
Sales	Compensation	683,879
	Supplies	193,378
	Travel	125,705
Total		$1,178,768

FIGURE 4.10 This is an example of a unidirectionally structured table that contains two sets of categorical items: departments and expense types.

Here's the bidirectional version:

Expense Type by Dept	Finance	Sales	Total
Compensation	160,383	683,879	844,262
Supplies	5,038	193,378	198,416
Travel	10,385	125,705	136,090
Total	$175,806	$1,002,962	$1,178,768

FIGURE 4.11 This is an example of a bidirectional table that contains the same information as in *Figure 4.10* above but uses less space despite the fact that it displays additional values.

Count the cells (i.e., the individual rectangles formed by the grid lines) in each table. You'll find that the unidirectional table contains a grid of three columns by eight rows, totaling 24 cells, and the bidirectional table contains a grid of four columns by five rows, totaling 20 cells. That doesn't seem like much of a difference until you notice that the bidirectional version actually contains totals for each expense type in the right column that are not included in the unidirectional version, so not only is it smaller, it contains more information.

Table Design Solutions

As we explored the structural variations of tables, you might have noticed an affinity between the types of relationships that can be displayed in tables and the two primary ways to structure tables. This affinity indeed exists and is worth noting. Below is a table that lists the five types of relationships down the rows and the two structural variations across the columns, with blank cells intersecting these rows and columns. Take a few minutes to think about each type of relationship and fill in the blanks with either Yes or No to indicate whether the structural variation can effectively display the corresponding relationship. If the structural type will only work in certain circumstances, note that as well, and describe the circumstances. You might find it helpful to construct sample tables and arrange them in various ways to explore the possibilities.

Relationship	Unidirectional	Bidirectional
Between a single set of quantitative values and a single set of categorical items		
Between a single set of quantitative values and the intersection of multiple categories		
Between a single set of quantitative values and the intersection of multiple hierarchical categories		
Among a single set of quantitative values associated with multiple categorical items		
Among distinct sets of quantitative values associated with a single categorical item		

· · · · · · ·

I'm confident that you've used your developing design skills effectively to identify solutions in the exercise above, but just so you have a nice, neat reference, I've included one in the *Summary at a Glance* that follows.

Now you have a conceptual understanding of tables that includes what they are, when to use them rather than graphs, what they're used for, how they're structured, and which type of structure works for particular quantitative relationships. Later, in Chapter 8, *Table Design*, we'll examine the specific design choices available to express quantitative messages in tables as clearly and powerfully as possible.

Summary at a Glance

Quantitative-to-Categorical Relationships

	Structural Type	
Relationship	**Unidirectional**	**Bidirectional**
Between a single set of quantitative values and a single set of categorical items	**Yes**	Not applicable because there is only one set of categorical items
Between a single set of quantitative values and the intersection of multiple categories	**Yes.** Sometimes this structure is preferable because of convention.	**Yes.** This structure saves space.
Between a single set of quantitative values and the intersection of multiple hierarchical categories	**Yes.** This structure can clearly display the hierarchical relationship by placing the separate levels of the hierarchy side by side in adjacent columns.	Yes. However, this structure does not display the hierarchy as clearly if its separate levels are split between the columns and rows.

Quantitative-to-Quantitative Relationships

	Structural Type	
Relationship	**Unidirectional**	**Bidirectional**
Among a single set of quantitative values associated with multiple categorical items	**Yes**	**Yes.** This structure works especially well because the quantitative values are arranged closely together for easy comparison.
Among distinct sets of quantitative values associated with a single categorical item	**Yes**	Yes. However, this structure tends to get messy as you add multiple sets of quantitative values.

5   VISUAL PERCEPTION AND GRAPHICAL COMMUNICATION

Because graphical communication is visual, it must express information in ways that human eyes can perceive and brains can understand. Our eyes and the parts of the brain that handle input from them work in particular ways. Thanks to science, how we see is now fairly well understood, from the initial information-carrying rays of light that enter our eyes to the interpretation of that information in the gray folds of the visual cortex. By understanding visual perception and its application to the graphical communication of quantitative information, you will learn what works, what doesn't, and why. This chapter brings the principles of graphical design for communication alive in ways that are practical and can be applied skillfully to real-world challenges in presenting quantitative information.

Vision, of all the senses, is our most powerful and efficient channel for receiving information from the world around us. Approximately 70% of the sense receptors in our bodies are dedicated to vision.

70%

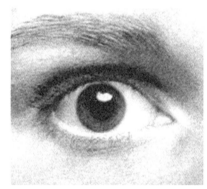

30%

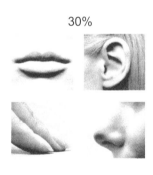

FIGURE 5.1 70% of the body's sense receptors reside in the retinas of our eyes.

There is an intimate connection among seeing, thinking, and understanding. It is no accident that the terms we most often use to describe understanding are visual in nature, such as insight, illumination, and enlightenment. "I see" is the expression we use when something begins to make sense. We call people of extraordinary wisdom seers.

Graphs, and, to a lesser extent, tables, are visual means of communication. Although tables are visual in that we receive their information through our eyes, their reliance on verbal language in the form of written text reduces the degree to which they take advantage of the power of visual perception. Nevertheless, inattention to the visual design of tables, such as improper alignment of numbers and excessive use of lines and fill colors, can greatly diminish their effectiveness. Graphs, in contrast, are primarily visual in nature. The power of graphs and the power of visual perception are intimately linked.

Equipped with a fundamental understanding of visual perception—how it works, what works, what doesn't, and why—your understanding of table and graph design will deepen beyond a memorized list of design principles. You will be able to apply these insights to new design challenges that extend well beyond the examples found in this book.

Colin Ware has written two wonderful books that explain how visual perception works and how this knowledge can be applied to graphical communication. Here's how he expresses the significance of visual perception:

> *Why should we be interested in visualization? . . . The visual system has its own rules. We can easily see patterns presented in certain ways, but if they are presented in other ways, they become invisible . . . When data is presented in certain ways, the patterns can be readily perceived. If we can understand how perception works, our knowledge can be translated into rules for displaying information. Following perception-based rules, we can present our data in such a way that the important and informative patterns stand out. If we disobey the rules, our data will be incomprehensible or misleading.*[1]

1. Colin Ware (2004) *Information Visualization: Perception for Design*, Second Edition. Morgan Kaufmann Publishers, page xxi.

The rules that the visual system follows are often not obvious. In the following figure, one set of circles appears to be bulging outwards and the other appears to be dented inwards. Which is bulging outwards?

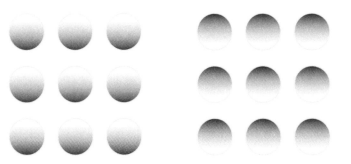

FIGURE 5.2 One set of circles appears to be bulging outwards and the other is dented inwards.

The one on the left—correct? Now I'm going to make a change and ask the question again. Which set is bulging outwards now?

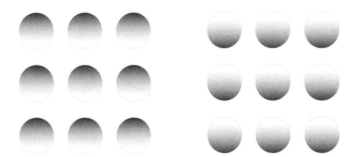

FIGURE 5.3 The set that appears to be bulging outwards has now switched positions.

Now the circles on the right appear to be bulging outwards, but I did not switch the position of the two sets; I merely turned the circles upside down. There is a reason why those on the right appear to bulge outwards: built into visual perception is the assumption that the light is coming from above. Because it looks as if light is reflecting off of their upper halves and shadows appear on their lower halves, the circles on the right are naturally perceived as bulging outwards. Until the invention of fire and more recently the invention of electric lights, all significant sources of illumination came from the sun, so light almost always came from above. Now that other sources of light are commonplace, perhaps this assumption will eventually disappear from the wiring in our brains, but for now it guides our perceptions. I mention this only to point out that rules are built into visual perception that aren't always obvious. We need to understand these rules if we want to present data in ways that work for people's eyes and make sense to their brains.

Mechanics of Sight

The process of seeing and making sense of what we see begins with sensation in our eyes and progresses to perception in our brains. Sensation is a physical process that occurs when a stimulus is detected. Perception is a cognitive process, involving the interpretation of the physical stimulus to make sense of it. For a visual stimulus to occur, there must be light. To see something, there must be an object out there that is either itself a source of light (e.g., the sun, a light bulb, a computer screen, a firefly) or a surface that reflects light (e.g., an apple, a piece of paper on which a table or graph is printed).

The light enters our eyes, which contain nerve cells specifically designed to absorb light and translate it into neural signals (the language of the nervous system). These neural signals are then passed via the optic nerve to our brains where they are processed in a series of stages to make sense out of what we've seen.

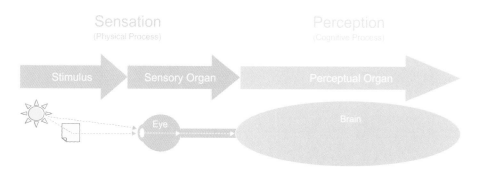

FIGURE 5.4 This diagram represents the stages in the process of visual perception.

The Eye

Our eyes operate a lot like cameras, with several components working together to sense information that is contained in light. The surface of the eye, called the *cornea*, is a protective covering. Inside the eye, behind the cornea, resides the *iris*, a muscle that works like a camera shutter, enlarging or decreasing the size of the opening, called the *pupil*, to allow just the right amount of light to enter. The pupil becomes small when the amount of light is high (i.e., bright) and enlarges when there is less light, thereby allowing enough light through to provide a discernable image. Behind the iris is a lens, similar to the lens of a camera, which thickens to focus on objects that are near and thins to focus on distant objects.

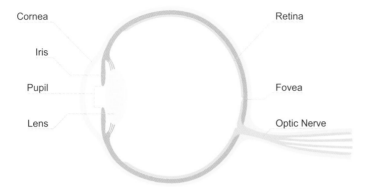

Cornea

Iris

Pupil

Lens

Retina

Fovea

Optic Nerve

FIGURE 5.5 This is a diagram of the eye, with a label for each of the components that are significant to vision.

Eye diagram created by Keith Stevenson.

Once light passes through the pupil, it shines on the rear inner surface of the eye, called the *retina*. This is a thin surface coated with millions of specialized nerve cells called *rods* and *cones*. Rods and cones differ in photosensitivy: rods specialize in sensing dim light and record what they detect in black and white, and cones specialize in sensing brighter light and record what they detect in color. Cones are divided into three types, each of which specializes in detecting a different range of the color spectrum: blue, green, or red.

In the center of the retina is a small area called the *fovea* where cones are densely concentrated. Images that are focused on the fovea are seen with extreme clarity because of the high number of cones there. When we want to examine something closely, we focus on it directly so that the light coming from it is directed to the fovea. When we focus in this manner on a printed page, we are able to see an amazing amount of detail, making distinctions between as many as 625 separate points of ink per square inch.[2]

Rods and cones encode what they detect into neural signals: electro-chemical pulses, the language of the nervous system. These signals are then passed along the optic nerve to the brain where perceptual processing occurs. Millions of cones and rods work together simultaneously, each sending signals to the brain at the same time through an extremely high-bandwidth channel at amazing speeds. Visual perception is the most highly developed and efficient of our senses.

The process of seeing is not smooth as you might imagine but consists of spurts of information interrupted by frequent, brief pauses. This is because the fovea can only focus on a limited area at a time. Our eyes fixate on a particular spot and remain there for up to half a second, taking in the details while at the same time we use what we are seeing in the non-foveal parts of the retina to survey the full field of vision and select our next point of concentration. It takes our eyes about a quarter second to jump to the next point of fixation where the process is repeated. These little jumps are called *saccades*, and the motion is called *saccadic eye movement*.[3]

As you are reading this page, your eyes are quickly jumping and fixating, jumping and fixating, taking in a few words at a time, moving from left to right across each line, then down to the next. During the brief moment when a saccade is occurring, when your eyes are jumping from one point of fixation to the next, you are not as sensitive to visual sensation as you are when your eyes

2. Edward Tufte (2001) *The Visual Display of Quantitative Information*, Second Edition. Graphics Press, page 161.

3. Much of the information about the saccadic nature of vision was derived from Marlana Coe (1996) *Human Factors for Technical Communicators*. John Wiley & Sons, Inc.

are not moving. If something popped up on your computer screen during a saccadic eye movement and disappeared before the movement was complete, you would probably never notice that it was ever there.

Despite the millions of bits of information that our eyes process every second, we only see a little of what is out there, and only a fraction of that in detail. Some things grab our attention more than others.

The Brain

We don't see images with our eyes; we see them with our brains. Our eyes record light and translate it into electro-chemical signals, passing them on to our brains, which is where images are actually perceived. What goes on in the brain produces what we experience as visual perception.

Although the analogy is less than perfect, it is useful to think of the brain as a computer. Computers involve components and processes that are like the components and processes of our brains. Computers receive input from a variety of devices (a keyboard, mouse, digital camera, etc.). Let's trace the path of input from the keyboard as an example of how computers process input from the outside world, which is comparable to the sensory information we perceive. When we tap the keys of a keyboard, electrical signals are initially routed into a special kind of memory called a *buffer*, in this case the keyboard buffer. This temporary location serves as a waiting room for the fraction of a second that the computer is busy until it is ready for fresh input. Once the computer is ready for the next input, these electrical signals move along the path into *working memory*, also called *random access memory (RAM)*. This is an extremely high-speed form of memory that the computer uses to process the information, doing what is necessary to interpret the message and respond to it. In RAM, bits and bytes are translated into meaningful terms, such as commands to take a particular action or words of a document. If the information that the computer has received is meant to be kept around for future reference, it is then moved into some form of permanent memory, such as a hard disk. These three types of memory—buffer memory, working memory, and permanent memory—are analogous to three types of memory that process visual information in our brains, which are called:

- *Iconic memory*
- *Working memory*
- *Long-term memory*

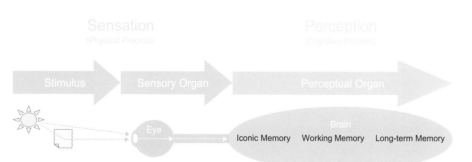

FIGURE 5.6 This is a diagram of visual perception, including the three types of memory that handle visual processing in the brain: iconic, working, and long-term memory.

ICONIC MEMORY

Signals from the eyes are initially routed through the optic nerve into iconic memory (a.k.a. the *visual sensory register*) where each snapshot of input waits to be passed on to working memory. Information ordinarily remains in iconic memory for less than a second while extremely rapid processing takes place before the information is forwarded to working memory. This part of perceptual processing is automatic and unconscious. For this reason it is called *preattentive processing*, as opposed to the conscious, higher-level cognitive processing that occurs later in working memory, which is called *attentive processing*. Preattentive processing is an extremely fast process of recognition and nothing more. Attentive processing, on the other hand, is a sequential process that takes more time and can result in learning, understanding, and remembering. Preattentive visual processing detects several attributes, such as color and the location of objects in 2-D space. Because preattentive processing is tuned to these attributes, they jump out at us and are therefore extremely powerful aspects of visual perception. If you want something to stand out as important in a table or graph, you should encode it using a preattentive attribute that contrasts with the surrounding information, such as red text in the midst of black text. If you want a particular set of objects to be seen as a group, you can do this by assigning them the same preattentive attribute.

Preattentive attributes play an important role in visual design. They can be used to distinctly group and highlight objects. We'll examine them in detail later in this chapter with special attention to how we can use them to design tables and graphs that communicate effectively.

WORKING MEMORY

As visual information is moved from iconic memory into working memory, what our brains deem useful is combined into meaningful visual *chunks*. These chunks of visual information are used by the conscious, attentive process of perception that occurs in working memory. Two fundamental characteristics of working memory are:

- It is temporary.
- It has limited storage capacity.

Information remains in working memory from only a few seconds to as long as a few hours if it is periodically rehearsed. Without rehearsal, it soon vanishes. With the right kind of rehearsal, it is stored in long-term memory where it remains for later use.

Only three or four chunks of information can be stored in working memory at any one time. For something new to be brought into working memory, something that is already there must either be moved into long-term memory or forgotten. Similar to RAM in a computer, working memory in the brain stores information temporarily for high-speed processing. As this processing occurs, new sensations or memories are being moved into working memory while others are either moved into long-term memory or forgotten.

Colin Ware points out, in *Information Visualization: Perception for Design*, that our brains appear to have more than one type of working memory; each kind specializes in a different type of information processing. For instance, research has demonstrated that there are separate areas for visual and verbal information and that these areas don't compete with each other for resources.

An important fact to keep in mind about working memory is that readers of tables and graphs can only hold a few chunks of information in their heads at any one time. For instance, if you design a graph that includes a legend with a different color or symbol shape for 10 different sets of data, your readers will be forced to constantly refer back to the legend to remind themselves which is which because working memory is limited to three or four concurrent chunks of information. Chunks of information can vary in size. By designing a visual display of information to form larger, coherent patterns that combine multiple data values into chunks, you can make it easier for your readers to hold more information in working memory. This is one of the reasons that graphs are capable of communicating a great deal of information that can be perceived all at once while tables are limited to use for looking things up. We can't take a bunch of numbers from a table and chunk them together meaningfully for storage in working memory; we can, however, discern in a graph the image of a single, meaningful pattern that consists of thousands of values.

LONG-TERM MEMORY

When we decide to store information for later use, either consciously or unconsciously, we rehearse that chunk of information to move it from working to long-term memory. How we store that information involves an intricate neural network of links and cross-references (i.e., associations), like indexes in a computer that help us to find information and retrieve it from a hard disk into working memory when we need it. One piece of information can have many associations.

Long-term memory is vitally important to visual perception because it holds our ability to recognize images and detect meaningful patterns, but we don't have to understand very much about long-term memory to become better designers of tables and graphs. Mostly, we need to know how to use preattentive visual attributes to grab and direct our readers' attention and how to work within the limits of working memory.

Evolution of Visual Perception

Like all other products of evolution, visual perception developed as an aid to survival. Because of its evolutionary roots, visual perception is fundamentally oriented toward action, always looking for what we can do with the things we see. As designers of tables and graphs, we should understand how to use attributes of visual perception to encode information in ways that others can easily, accurately, and efficiently decode.

Attributes of Preattentive Processing

Preattentive processing occurs below the level of consciousness at an extremely high speed and is tuned to detect a specific set of visual attributes. Attentive processing is conscious, sequential, and much slower. The difference between preattentive processing and attentive processing is easy to demonstrate. Take a look at the four rows of numbers below and determine, as quickly as you can, how many times the number 5 appears in the list:

987349790275647902894728624092406037070570279072
803208029007302501270237008374082078720272007083
247802602703793775709707377970667462097094702780
927979709723097230979592750927279798734972608027

FIGURE 5.7 This example demonstrates the slow speed at which we perceive information using a serial process, one number at a time.

How many? The answer is six. Even if you got the right answer (many don't when I do this exercise in class), it took you several seconds to perform this task because it involved attentive processing. The list of numbers did not include any preattentive attributes that could be used to distinguish the fives from the other numbers, so you were forced to perform a sequential search looking for the specific shape of the number five. Now do it again, this time using the list below:

987349790275647902894728624092406037070570279072
803208029007302501270237008374082078720272007083
247802602703793775709707377970667462097094702780
927979709723097230979592750927279798734972608027

FIGURE 5.8 This example demonstrates the fast speed with which we process visual stimuli that exhibit preattentive attributes.

This time it was easy because the fives were distinguished through the preattentive attribute of color intensity. Because only the fives are black, and all the other numbers are light gray, the black fives stand out in contrast to the rest. This example shows the power of preattentive attributes used knowledgeably for visual communication.

Colin Ware has organized preattentive attributes into four categories: form, color, spatial position, and motion. I've reduced his list of attributes somewhat and removed the category of motion entirely, to focus on attributes that we find most relevant to static tables and graphs:

Colin Ware (2004) *Information Visualization: Perception for Design*, Second Edition. Morgan Kaufmann Publishers, pages 151 and 152

Category	Attribute
Form	Length
	Width
	Orientation
	Shape
	Size
	Enclosure
Color	Hue
	Intensity
Spatial position	2-D position

In each illustration of a preattentive attribute on the following page, only one object stands out as different from the rest, based on a variation of the attribute.

Attributes of Form

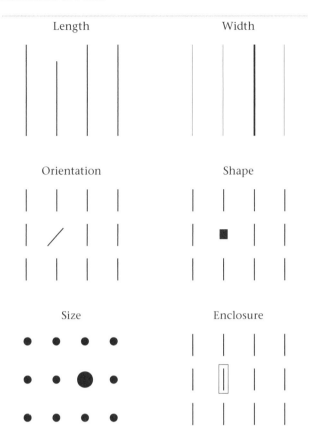

For our purposes, length is distance along the dominant dimension of an object. For example, the lines in the illustration above primarily vary in the vertical dimension only. Think in terms of bars in a bar graph, which encode quantitative values as length along a single dimension—either vertically or horizontally, but not both. Whenever we use variation in length to encode values in graphs, the objects, such as bars, will share a common baseline, which makes it extremely easy to compare their lengths. By width, we're referring to variation along the secondary dimension of an object. Once again, think of bars that vary in one dimension, such as vertical length, to encode values, but also vary in their less dominant dimension, width, to encode a second set of values.

Attributes of Color

Hue is a precise term for what we normally think of as color (red, green, blue, pink, etc.). Color is made up of three separate attributes; hue is one. *Intensity* applies to both of the other two attributes of color: *saturation* and *lightness*. In the previous illustration only intensity varies in that one object is lighter than the others.

The primary system that is used to describe color is known as the *HSL* (hue, saturation, and lightness) system. This system describes hue numerically in degrees from 0 to 360 along the circumference of a circle, illustrated by the following color wheel:

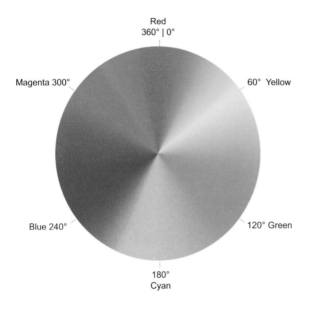

FIGURE 5.9 This is a color wheel, which represents hue in a circular manner.

Numbering starts at the top with red at 0° and continues around the wheel until it reaches the point where it started with red again, which is also assigned the value of 360°. If you could walk around this color wheel clockwise taking 60° steps, starting at red (0°), with each succeeding step you would be positioned at the following hues: yellow (60°), green (120°), cyan (180°), blue (240°), magenta (300°), then back to red again (360°).

Saturation measures the degree to which a particular hue fully exhibits its essence. In the example below, red ranges from completely pale on the left, lacking the essence of red entirely, to fully red on the right, exhibiting its full essence. Saturation is expressed as a percentage, starting with 0% on the left, which represents no saturation, to 100% on the right, which represents full saturation.

FIGURE 5.10 This is an illustration of the full range of saturation for the hue red, with 0% saturation on the left and 100% saturation on the right.

Lightness (also known as brightness) measures the degree to which a color appears dark or bright. In the next example, red ranges from fully dark (black) on the left to bright red on the right. Lightness is also expressed as a percentage, with 0% representing fully dark and 100% representing fully light.

0% lightness   100% lightness

FIGURE 5.11 This is an illustration of the full range of lightness for the hue red, with 0% lightness (dark) on the left and 100% lightness (bright) on the right.

Any color can be described numerically based on these three measures: hue (0 – 360°), saturation (0 – 100%) and lightness (0 – 100%).

Attributes of Spatial Position

2-D Position

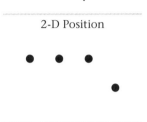

Our perception of space is primarily two dimensional. We perceive differences in vertical position (up and down) and in horizontal position (left and right) clearly and accurately. We also perceive a third dimension, depth, but not nearly as well.

Applying Visual Attributes to Design

Preattentive attributes can be used to make particular aspects of what we see stand out from the rest. Understanding these attributes enables us to design tables and graphs that visually emphasize the most important information they contain, keeping the other visual elements from competing for attention with key information.

3-D position is also a preattentive attribute, but it cannot be applied effectively to graphs on a flat surface (e.g., a page or computer screen), which relies on illusory cues of light and shadow, occlusion, and size to simulate depth. Even if 3-D position could be used effectively on a flat surface, it is not an effective attribute to use in table and graph design because our perception of depth is weak compared to our perception of height and width.

Uses for Encoding Quantitative Values

Some attributes of preattentive processing can be expressed as values along a continuum. For instance, the preattentive attribute of length can be expressed as a line of any length that the eye can see, not simply as short and long. We perceive quantitative meanings in the variations of some visual attributes but not in others. Using the attribute of length once again as an example, we naturally perceive long lines as greater than short lines. In this sense, we perceive the differing values of length quantitatively. But what about an attribute like hue? Which represents the greater value: blue, green, red, yellow, brown, or purple? Although it is true that each of these hues can be measured by instruments to determine its wavelength, and the number assigned to one hue would be greater or less than the number assigned to another, we don't think in these terms when we perceive hue. We perceive different hues only as categorically different, not quantitatively different; one hue is not more or less than another, they're just different. In contrast, we perceive color intensity quantitatively, from low to high.

Whether we perceive visual attributes as quantitative or categorical is significant. How we perceive these attributes tells us which ones we can use to encode quantitative values in graphs and which can only be used to encode categorical differences. The following list indicates which preattentive attributes are perceived quantitatively:

Type	Attribute	Quantitatively Perceived?
Form	Length	Yes
	Width	Yes, but limited
	Orientation	No
	Size	Yes, but limited
	Shape	No
	Enclosure	No
Color	Hue	No
	Intensity	Yes, but limited
Position	2-D position	Yes

Although one might argue that we can also perceive orientation quantitatively, there is no natural association of greater or lesser value with this characteristic (e.g., which is greater, a vertical or a horizontal line?).

All of these preattentive attributes can be used to encode categorical differences, including those that we perceive quantitatively. The most effective of these is 2-D position. For instance, the 2-D position of data points in a graph can simultaneously represent quantitative values based on the position of the data point relative to the Y axis and categorical differences based on their position along the X axis, as in the following example:

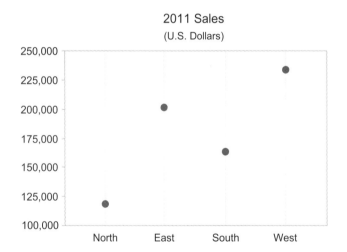

FIGURE 5.12 This graph uses the 2-D position of the data points to encode quantitative values along the Y axis and categorical items along the X axis.

Whenever you can use either length or 2-D position to encode quantitative values in graphs, you should do so rather than using other attributes that can be perceived quantitatively only to a limited degree. Even though we can tell that one value is greater than another when we use width, size, or color intensity, these attributes do not indicate precisely how much a value differs. This is true of width and size because both rely on our ability to assign values to the 2-D areas of objects, such as the slices of a pie chart or the width of a line. Although we can do a fairly good job of discerning that one object has a larger area than another, it is difficult to perceive by how much or to assign a value to the area. Take a look at the two circles on the following page. I've assigned a value to the one on the left, and it is up to you to assign a value to the one on the right.

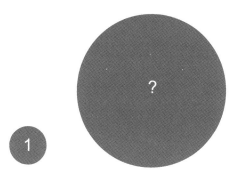

FIGURE 5.13 This figure illustrates the difficulty of perceiving differences in the 2-D areas of objects.

How much bigger is the large circle than the small circle? Given the fact that the size of the small circle equals a value of 1, what is the size of the large circle?

· · · · · · ·

The answer is 16. If you got the answer right, either you got lucky or you have rare perceptual talent. When I ask this question in a class of 70 students, answers typically range from 5 to 50.

Color intensity presents a similar problem. For instance, we can use color intensity along a gray scale ranging from white as the minimum value to black as the maximum. As long as there is enough difference in the color intensity of two objects, we can tell that one is darker than the other, but it is hard to assign a value to that difference. For this reason, it is best to avoid color intensity as a means of encoding quantitative value if the attributes of length or 2-D position are available to use instead.

Perceptual Effects of Context

One fact that may surprise you about our perception of visual attributes such as color or size is that our visual senses are not designed to perceive absolute values but rather differences in values. Your sense of the brightness of a candle will differ depending on the context, i.e., in relation to the other light in your range of vision. If you light a candle when surrounded by bright sunshine, the candle won't seem very bright. If you light the same candle while standing in a pitch-black cellar, the candle will seem extremely bright in contrast to the darkness. That's because our visual receptors measure differences rather than absolute values. Look at the four squares contained in the large rectangle below:

FIGURE 5.14 This illustration demonstrates the effect of context on the perception of color intensity.

How much darker is the square on the left than the square on the right?

· · · · · · ·

The truth is, there is no difference at all in the actual values of the four squares when measured by an instrument. All four are 50% gray, precisely in the middle between pure white and pure black. They appear different to us in this illustration, however, because our perception of them is influenced by the differences in the color intensity that surrounds them.

Every attribute of visual perception is influenced by context. In the following figure, the upper horizontal line looks longer than the lower, but they are exactly the same. The different shapes at the ends of the two horizontal lines affect our perception of their lengths

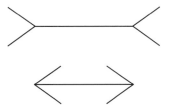

Optical illusions such as these clearly demonstrate the effect of context on visual perception. Even when we know we are looking at an optical illusion, misperception persists, because visual perception measures differences in the attributes of what we see, not absolute values. It's up to us to use this knowledge when designing tables and graphs to prevent context from causing a misperception of the message we're trying to communicate.

Some combinations of an object's visual attributes and its context can work well to highlight the object, but some can render it imperceptible. For instance, black text on a white background is easy to read. Other combinations are also acceptable, such as white text on a black background, and, to a lesser degree, red or dark blue text on a white background. In contrast, some combinations simply don't work, such as yellow on a white background or blue on a black background.

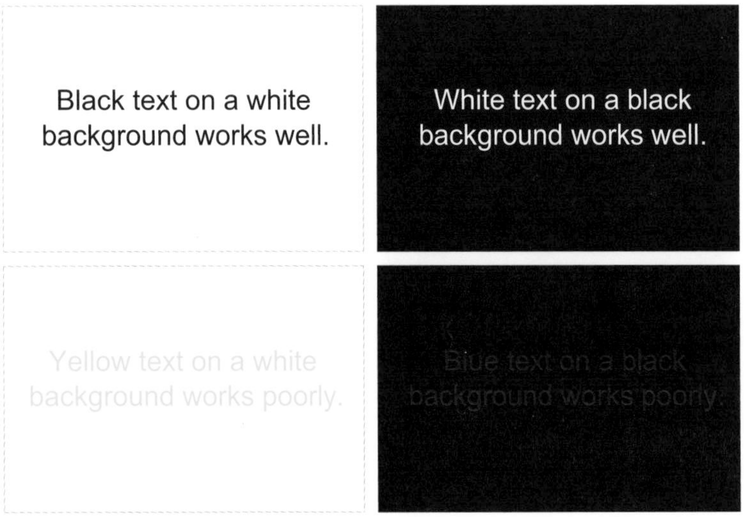

Black text on a white background works well.	White text on a black background works well.
Yellow text on a white background works poorly.	Blue text on a black background works poorly.

We could examine many more examples, both good and bad, but it's not necessary to memorize the combinations that do and don't work. As you can see

One practical application of the lesson that we learn from *Figure 5.14* is that we shouldn't use a gradient of fill color in the background of a graph. Doing so would cause objects (e.g., bars) on the left to look different from objects on the right even if they're exactly the same.

FIGURE 5.15 This figure demonstrates the effect of context on the perception of length.

FIGURE 5.16 Some contrasting hues work well for clear perception, and some don't. You must look closely to see that there is any text at all in the lower right rectangle because blue does not stand out distinctly against a black background.

from these examples, it's fairly easy to tell the difference when you take the time to look. Pay attention and use your own sense of visual perception to select combinations that are effective for clear and efficient communication and avoid those that aren't.

Limits to Distinct Perceptions

There is a limit to the number of expressions of a single attribute that can be clearly distinguished in a graph. For instance, if you create a line graph displaying departmental expenses across time, with a separate line of a distinct hue to represent each department in your company, readers of the graph will only be able to easily distinguish a limited number of hues. This is true for each visual attribute that can be used to encode categorical items.

Our ability to distinguish visual attributes diminishes as the number of alternatives increases. According to Ware:

> Pre-attentive symbols become less distinct as the variety of distracters increases. It is easy to spot a single hawk in a sky full of pigeons, but if the sky contains a greater variety of birds, the hawk will be more difficult to see. A number of studies have shown that the immediacy of any pre-attentive cue declines as the variety of alternative patterns increases, even if all the distracting patterns are individually distinct from the target.[3]

According to research, when reading graphs, we can distinguish preattentively between no more than about eight different hues, about four different orientations, and about four different sizes. All the other visual attributes of preattentive perception should be limited to a few values as well.[4] When we introduce a greater number of alternatives, people can no longer distinguish them preattentively and must therefore switch to slower attentive processing.

You might think that it would be worthwhile to use a larger number of distinct values even though this would force your readers make use of slower, attentive processing to interpret your graph. Unfortunately, if you do this, your readers will run up against the limits of working memory, which allows them to remember no more than four distinct values at a time.

Another limitation to keep in mind is that preattentive processing usually cannot handle more than one visual attribute of an object at a time. Let me illustrate. In the figure below, only the attribute of color intensity is used to distinguish between three values, represented with black, dark gray, and light gray:

3. Colin Ware (2004) *Information Visualization: Perception for Design*, Second Edition. Morgan Kaufmann Publishers, page 152.

4. Ibid. page 182

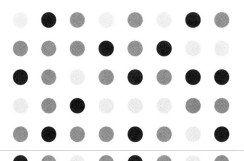

FIGURE 5.17 This figure illustrates the use of distinct values of a single visual attribute, in this case color intensity, to encode objects as different from one another.

It is not difficult to pick out the black circles as distinct from those that are dark gray or light gray, the dark gray circles as distinct from those that are black or light gray, and so on. Let's complicate it now by introducing a second preattentive attribute. By using two shapes (circles and squares), along with the three color intensities (black, dark gray, and light gray), we double the number of distinct values, but we do so at a cost. Take a look.

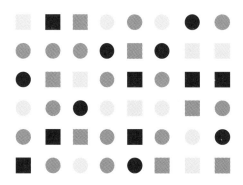

FIGURE 5.18 This figure illustrates the problem that results from encoding distinct values using more than one visual attribute, in this case color intensity and shape.

It's still fairly easy to focus on just the squares, or just the circles, or just the black objects, or just the dark gray objects, or just the light gray objects, but try to pick out the dark gray squares or the light gray circles. The process of simultaneously focusing on multiple visual attributes (e.g., shape and color intensity) requires slower, attentive processing. You can imagine how much more complicated it would get if we threw in a few more shapes or added the third attribute of size.

A related consideration is that when you select the expressions of a single attribute that you will use in a graph (e.g., multiple hues to distinguish categorical items), you must make sure that they are far enough apart from each other along the continuum of potential expressions to stand out clearly as distinct. Although I can choose up to four different shades of gray, shades that are too close to one another won't work. Here are two sets of four distinct values, encoded as different shades of gray. The four shades on the left were carefully chosen from the full range of potential values so they could easily be distinguished from one another, but those on the right were not.

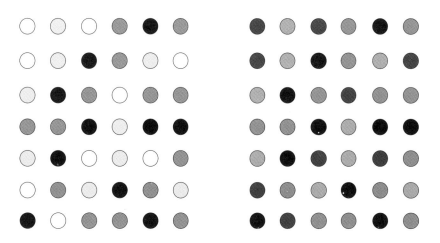

FIGURE 5.19 This figure illustrates the importance of selecting values of a visual attribute that can be easily distinguished. The four shades of gray encoded in the circles on the left are distinct, but the four on the right are not.

With a little effort, you can develop once and for all a set of distinct values for each of the visual attributes, and you can use this set over and over, rather than repeatedly going through the process of determining the distinctions that work.

Here's a list of nine hues that are easy to recognize and distinct enough to work well together:

1. Gray
2. Blue
3. Orange
4. Green
5. Pink
6. Brown
7. Purple
8. Yellow
9. Red

An excellent source of color combinations that work well together for data presentation can be found on the website www.ColorBrewer.org.

Each of these hues meets the requirement of distinctness, but the context of their use should sometimes lead you to select some and reject others. A good example, especially if you communicate with an international audience, involves cultural context. Particular hues carry meanings that vary significantly among cultures. In most western cultures, we think of red as signifying danger, warning, heat, and so on, but in China, red represents good fortune. The key is to be in touch with your audience and take the time to consider any possible associations with particular hues that might influence their interpretation of your message.

Various colors also affect us in different ways on a fundamental psycho-physical level. Some colors are strong and exciting, grabbing our attention, and others are more neutral and soothing, fading more into the background of our attention. Edward Tufte suggests that colors found predominantly in nature, especially those that are not overly vibrant (e.g., medium shades of gray, green, blue, and brown), are soothing and easy on our eyes.[5] Such colors are particularly useful in tables and graphs for anything that you don't want to stand out above the other content. On the other hand, fully saturated, bright versions of just about any hue will demand attention. These should only be used when you want to highlight particular information. Notice the difference between the medium shades on the left and the brighter shades on the right:

5. Edward R. Tufte (1990) *Envisioning Information.* Graphics Press, page 90.

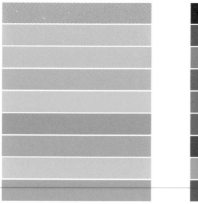 

FIGURE 5.20 This figure shows soft, soothing hues on the left, and bright, attention-getting hues on the right.

Another consideration when selecting a palette of colors is that our ability to distinguish colors decreases along with the sizes of objects. For this reason, small objects such as data points in graphs should be darker than large, visually weighty objects such as bars. In the two graphs below, the same colors are used in both. The colors of the bars in the upper graph are quite distinct, but they are somewhat difficult to distinguish in the lower graph because the points are small.

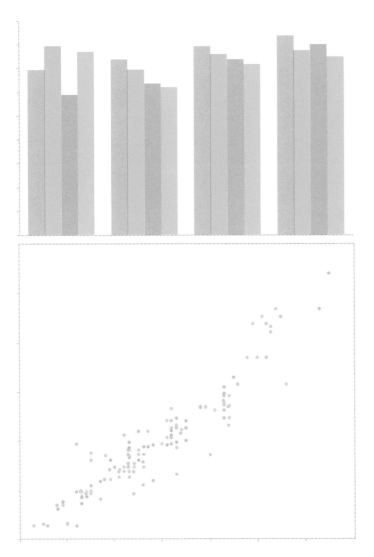

FIGURE 5.21 The same colors that allow us to easily distinguish bars are much harder to distinguish when used for small objects such as data points.

When selecting your palettes of colors for encoding categorical distinctions in graphs, it works best to maintain three versions: one with relatively light, yet distinct colors for large objects such as bars, one with darker versions of the same colors for small objects such as lines, and one with bright versions of the same colors for objects that must stand out above the rest. Here's an example of three versions of the nine distinct hues that were listed previously, two versions for ordinary use and one for highlighting:

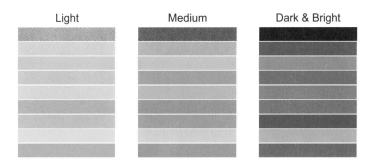

Light   Medium   Dark & Bright

FIGURE 5.22 These sample colors could be used for the three color palettes that you should maintain for displaying data. The RGB values of each are provided in Appendix I – *Useful Color Palettes.*

Another consideration when using hues to encode data is that about 10% of males and 1% of females suffer from colorblindness, unable to distinguish certain colors because of a lack of cones (color receptors) in the eye that detect a certain range of hue. Most people who are colorblind lack the cones that enable them to distinguish between red and green so that, to these viewers, both colors appear brown. When your audience is likely to include people who are colorblind, it is best to use only red or green, but not both, unless you are careful to vary their intensities sufficiently so that they appear different from one another.

Limits to the Use of Contrast

Visual perception evolved in a way that makes us particularly aware of differences, i.e., features that stand out in contrast to the rest. When we look at something, our eyes are not only drawn to differences, but our brains insist on making some sense of those differences. If you create a table of information and all the text is black except for one column that is red, readers will assume that the difference is significant and that the red text is especially important. It is up to us to only include visual differences that are meaningful.

Contrast is most effective when only one thing is different in a context of other things that are the same. As the number of differences increases, the degree to which those differences stand out decreases. In the graph below, the attributes were controlled so that the bar representing "Contributions from People like You" would clearly stand out from the rest because of a single difference: its color.

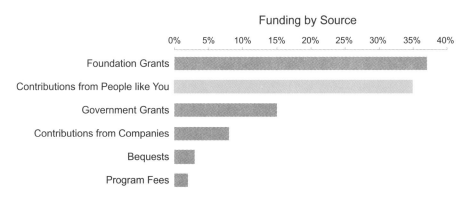

Funding by Source

FIGURE 5.23 This graph uses contrast effectively to highlight the most important information.

If you try to emphasize so many things that emphasis becomes the rule rather than the exception, your message will become buried in visual clutter. When everyone in the room shouts, no one can be heard.

Gestalt Principles of Visual Perception

Many fields of scientific study have contributed to our understanding of visual perception, but none have revealed more that is relevant to visual design than the Gestalt School of Psychology. The original intent of Gestalt research when it began in 1912 was to uncover how we perceive pattern, form, and organization in what we see. The German word *gestalt*, however lofty it may seem to speakers of English, simply means "pattern." The founders observed that we organize what we see in particular ways in an effort to make sense of it. The result of their research was a series of Gestalt principles of perception, which are still respected today as accurate descriptions of visual behavior. These principles reveal that we tend to group objects in particular ways. Many of these findings are relevant to our interest in the design of tables and graphs as well as to the overall design of the reports that contain them. This section is an introduction to Gestalt principles.

Principle of Proximity

We perceive objects that are close to each other as belonging to a group. In the following illustration, we naturally perceive the 10 circles as three groups because of the way they are spatially arranged:

FIGURE 5.24 This figure illustrates the Gestalt principle of proximity.

This principle can be applied to the design of tables. We can direct our readers to scan predominantly across rows or down columns of data by spacing the data to accentuate either the row or the column groupings, as illustrated below:

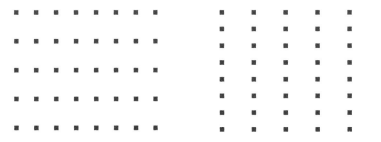

FIGURE 5.25 This figure illustrates an application of the Gestalt principle of proximity to table design.

In response to subtle grouping cues, we are naturally inclined to scan the squares on the left by row and those on the right by column. This is because the squares on the left are slightly closer together horizontally than vertically, which groups them by rows, and the squares on the right are slightly closer together vertically than horizontally, which groups them by columns. The conscious use of space to group objects together or keep them separate is a powerful tool for organizing information and directing your readers to view it in a particular way.

Principle of Similarity

We tend to group together objects that are similar in color, size, shape, or orientation. Here are some examples:

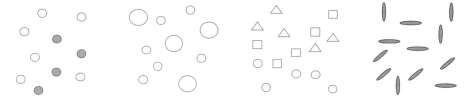

FIGURE 5.26 These examples Illustrate the Gestalt principle of similarity.

This principle reinforces what we've already learned about the usefulness of color (both hue and intensity), size, shape, and orientation to display categorical groupings. When these attributes are used to encode a distinction between different sets of data in a graph, they work quite effectively as long as the number of distinctions is kept to a few and the attributes are different enough from one another so that we can clearly distinguish the groups. We can use this principle to direct readers of tables to focus primarily across rows or down columns, as in the following example:

XXXXX	XXXXX	**XXXXX**	XXXXX	**XXXXX**	XXXXX	**XXXXX**	XXXXX
XXXXX	XXXXX	**XXXXX**	XXXXX	**XXXXX**	XXXXX	**XXXXX**	XXXXX
XXXXX	XXXXX	**XXXXX**	XXXXX	**XXXXX**	XXXXX	**XXXXX**	XXXXX
XXXXX	XXXXX	**XXXXX**	XXXXX	**XXXXX**	XXXXX	**XXXXX**	XXXXX
XXXXX	XXXXX	**XXXXX**	XXXXX	**XXXXX**	XXXXX	**XXXXX**	XXXXX
XXXXX	XXXXX	**XXXXX**	XXXXX	**XXXXX**	XXXXX	**XXXXX**	XXXXX
XXXXX	XXXXX	**XXXXX**	XXXXX	**XXXXX**	XXXXX	**XXXXX**	XXXXX
XXXXX	XXXXX	**XXXXX**	XXXXX	**XXXXX**	XXXXX	**XXXXX**	XXXXX
XXXXX	XXXXX	**XXXXX**	XXXXX	**XXXXX**	XXXXX	**XXXXX**	XXXXX

FIGURE 5.27 This figure demonstrates the use of the principle of similarity to arrange data column-wise in a table.

Principle of Enclosure

We perceive objects as belonging together when they are enclosed in a way that appears to create a boundary around them (e.g., a border or common field of color). They appear to be set apart in a region that is distinct from the rest of what we see. Notice how strongly your eyes are led to group the enclosed objects below:

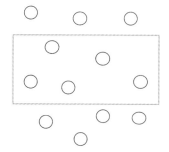 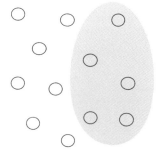

FIGURE 5.28 These two examples illustrate the Gestalt principle of enclosure; in the left-hand example, the enclosure is formed by a border, and, in the right-hand example, by a gray region.

The two sets of circles are arranged in precisely the same way, yet the differing enclosures lead us to group the circles differently. This principle is exhibited frequently in tables and graphs by the use of borders and background fill colors to group information and sometimes to emphasize particular items. As you can see in the previous illustration, a strong enclosure (e.g., bright, thick lines or dominant colors) is not necessary to create a strong perception of grouping. In fact, we sometimes perceive enclosures that aren't really there, such as the white shape that you see in the middle of the gray objects below:

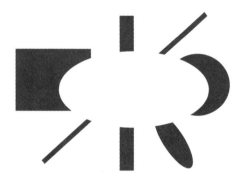

FIGURE 5.29 This is an example of an illusory contour (i.e., perception of a white oval in the middle of the gray objects, forming a separate region of white space).

The reason that we perceive this illusory enclosure is explained by the next two Gestalt principles: the principle of closure and the principle of continuity.

Principle of Closure

We have a keen dislike for loose ends. We will always try to resolve ambiguous visual stimuli to eliminate the ambiguity. So if we are faced with objects that could be perceived on the one hand as open, incomplete, and unusual forms; or on the other hand as closed, whole, and regular forms, we will naturally perceive them as the latter. In the illusory enclosure illustrated above, it is natural to perceive the gray objects as rectangles, circles, ovals, and continuing lines, partially hidden by a white oval, rather than as a collection of oddly shaped objects. The principle of closure asserts that we perceive open structures as closed, complete, and regular (e.g., the gray objects above) whenever there is a way that we can reasonably interpret them as such. Here's a more direct illustration of this principle:

FIGURE 5.30 This figure illustrates the principle of closure.

It is natural for us to perceive the lines on the left as a rectangle rather than as two sets of three connected and perpendicular lines and to perceive the object on the right as an oval rather than a curved line.

We can put this tendency to perceive whole structures to use when designing tables and graphs. For example, we can group objects (e.g., points, lines, or bars on a graph) into visual regions without the use of complete borders to define the space. It is not only sufficient but usually preferable to define the area of a graph through the use of a single set of subtle X and Y axes rather than with heavy lines and fill colors that enclose it completely, as in the following examples:

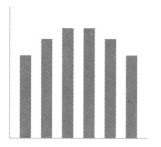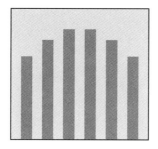

FIGURE 5.31 These examples demonstrate an application of the principle of closure. We can define the area of a graph with subtle X and Y axes, as seen on the left, thus minimizing the visual weight of a graph's supporting components in relation to its data components.

Principle of Continuity

We perceive objects as belonging together, as part of a single whole, if they are aligned with one another or appear to form a continuation of one another. In the left-hand illustration below, we see the various visual objects as forming a simple image of a rectangle and a wavy line. If we separated the two objects, we assume they would look like those in the middle illustration. We don't see a rectangle and three curved lines, like those in the right-hand illustration even though this is another possible interpretation of the image. Our tendency is to see the shapes as continuous to the greatest degree possible.

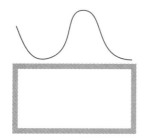

FIGURE 5.32 These examples illustrate the principle of continuity.

In the bar graph below, the left alignment of the bars makes it obvious that they share the same baseline, thus eliminating the need to reinforce the fact with a vertical axis:

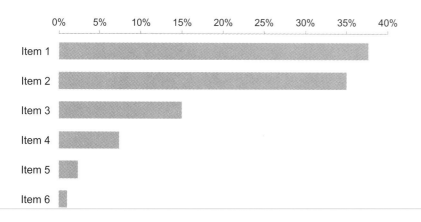

FIGURE 5.33 This is an example of applying the principle of continuity to graph design.

In the table below, it is obvious which categorical items are division names and which are department names, based on their distinct alignments:

Division	Department	Headcount
G&A	Finance	15
	Purchasing	5
	Information Systems	17
Sales	Field Sales	47
	Sales Operations	10
Engineering	Product Development	22
	Product Marketing	5

FIGURE 5.34 This is an example of applying the principle of continuity to table design.

The left alignment of the divisions and departments and the right alignment of the headcounts creates a strong sense of grouping without any need for vertical grid lines to delineate the columns. Even if two columns of data overlap, with no white space between them, as long as they are separately aligned, they remain clearly distinguishable, as in the following example:

Division/Department	Headcount
G&A	
Finance	15
Purchasing	5
Information Systems	17
Sales	
Field Sales	47
Sales Operations	10
Engineering	
Product Development	22
Product Marketing	5

FIGURE 5.35 This is an example of applying the principle of continuity to table design to save horizontal space.

In this example, the principle of continuity was used to align related categories in a way that saved horizontal space.

Principle of Connection

We perceive objects that are connected (e.g., by a line) as part of the same group. In the example below, even though the circles are equally spaced from one another, the lines that connect them create a clear perception of two vertically attached pairs:

Technically, the principle of connection was not directly presented by the original Gestalt School of Psychology but has been articulated since as an extension of the principle of continuity.

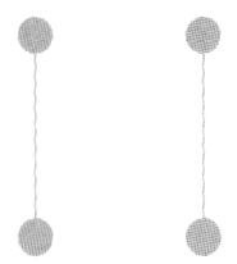

FIGURE 5.36 This figure illustrates the principle of connection.

Connection exercises even greater power over visual perception than proximity or similarity (color, size, and shape) but less than enclosure, as you can see from the examples below:

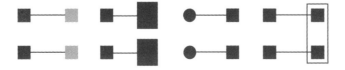

FIGURE 5.37 These are examples of the relative perceptual strength of the Gestalt principles that we've examined so far.

The perceptual strength of connection makes lines a useful method of connecting items in graphs. Points without lines in graphs are much harder for our eyes to connect. Notice the difference between the two examples below:

FIGURE 5.38 These examples illustrate the power of lines to reveal connectedness.

Lines in a graph not only create a clear sense of connection but also bring to light the overall shape of the data, when appropriate, which couldn't be discerned without them.

Summary at a Glance

Mechanics of Sight

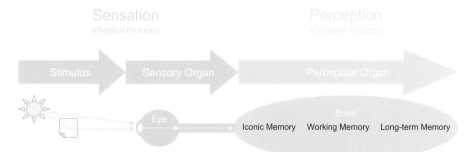

Attributes of Preattentive Processing

Type	Attribute	Quantitatively Perceived?
Form	Length	Yes
	Width	Yes, but limited
	Orientation	No
	Size	Yes, but limited
	Shape	No
	Enclosure	No
Color	Hue	No
	Intensity	Yes, but limited
Position	2-D position	Yes

Gestalt Principles of Visual Perception

Principle	Description
Proximity	Objects that are close together are perceived as a group.
Similarity	Objects that share similar attributes (e.g., color or shape) are perceived as a group.
Enclosure	Objects that appear to have a boundary around them (e.g., formed by a border or area of common color) are perceived as a group.
Closure	Open structures are perceived as closed, complete, and regular whenever there is a way that they can be reasonably interpreted as such.
Continuity	Objects that are aligned together or appear to be a continuation of one another are perceived as a group.
Connection	Objects that are connected (e.g., by a line) are perceived as a group.

6   FUNDAMENTAL VARIATIONS OF GRAPHS

Different quantitative relationships require different types of graphs. This chapter explores the fundamental variations of graphs that correspond to different quantitative relationships and then pairs these variations with the visual components and techniques that can present them most effectively.

Graphs always feature one or more specific relationships between values. The strength of graphs is their exceptional ability to present complex relationships so that we can see and compare them quickly and easily. The visual nature of graphs imbues them with this power. In this chapter we will first examine the methods that graphs use to visually encode data, both quantitative and categorical; then, we'll identify each type of relationship that graphs can display, and we'll finish by exploring the structural forms that can be used most effectively to display each relationship.

Encoding Data in Graphs

Think of yourself as a craftsperson, such as a carpenter, in the midst of your apprenticeship. It's my job to introduce you to the craft, helping you learn the required knowledge and skills until they become second nature and you can do the work confidently on your own. At the moment, we are going to add a few simple tools to your tool belt and make sure that you know when and how to use them.

Graphs consist of several components (scales along axes, grid lines, bars, legends, etc.). Some components represent quantitative values (lines, bars, etc.), some represent categorical items, and some play a supporting role. The structural variations of graphs are defined primarily by differences in the components that encode quantitative values (e.g., lines versus bars).

Graphical Means for Encoding Quantitative Values

Quantitative values can be represented in graphs using any of the following:

- Points
- Lines
- Bars
- Boxes
- Shapes with varying 2-D areas
- Shapes with varying color intensity

These graphical encoding methods vary in their ability to display quantitative information effectively. Each has its strengths and limitations that make it well suited for some purposes and ill suited for others.

POINTS

By a *point*, I mean any simple and small geometric shape that is used to mark a specific location on a graph. A point often consists of a dot. In the following example of a scatter plot, a separate point encodes each quantitative value.

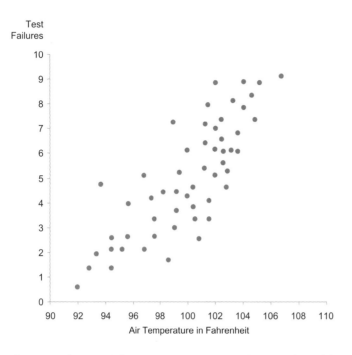

FIGURE 6.1 This graph uses points, in this case dots, to encode quantitative values.

The point closest to the top right represents an approximate value of 9 test failures on the Y axis and 107° on the X axis. In addition to a dot, any simple symbol (e.g., a small square or triangle) may be used to mark values on a graph.

Points are not necessarily restricted to scatter plots. Here's another example of a graph that uses points:

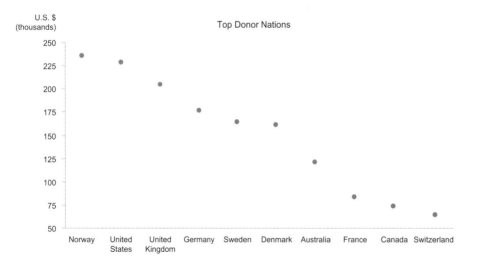

FIGURE 6.2 This graph uses points to encode quantitative values, this time in a form that is not a scatter plot.

When points are used in this manner to do what you might also use a bar graph to do—that is to display values along a nominal or ordinal categorical scale, such as this list of countries—the graph is called a *dot plot*. Why not use bars in this case? We'll get to that in a minute.

LINES

Lines are usually used to connect a series of values in a graph. In the example below, a line connects 12 monthly sales values. The position of each point along the line encodes each value in relation to the quantitative scale.

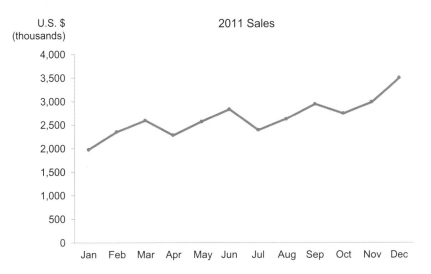

FIGURE 6.3 This graph employs two graphical objects, points and a line, to represent the same set of quantitative values.

Because the points are connected by a line, we can view the entire series of values as a single pattern from left to right. This is especially useful for displaying how values change through time because it is quite easy to think of the up and down patterns as degrees of change.

Lines are routinely used in graphs to represent values in two ways:

- To connect individual data values (as in the example above)
- To display the overall trend of a series of values (such as in the form of a trend line in a scatter plot)

When a line connects individual values, we do not always need to display those values as points along the line. Unlike the example in *Figure 6.3*, which combines points and a line, the following example uses only a line:

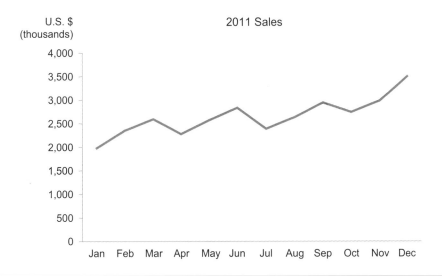

FIGURE 6.4 This graph uses only a line to encode the quantitative values.

A line that displays a trend is called a trend line or line of best fit. Here's an example of a trend line in a scatter plot:

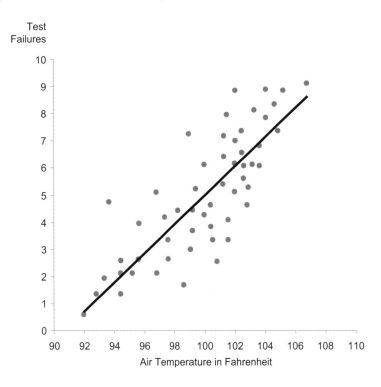

FIGURE 6.5 This graph uses points to represent individual quantitative values and a line to summarize the overall trend of those individual points.

Trend lines are used to summarize the overall pattern that appears in a set of values. In this particular scatter plot, the trend is linear (straight) and positive (sloped upward from left to right), which indicates that, as air temperature increases, so does the number of test failures. In a scatter plot, a trend line is drawn from left to right to trace a path through the middle of the set of values. Trends are not always straight. When the overall pattern is curved in some manner, it is proper to say that the graphical object that summarizes it is a curve rather than a line. Whether linear or curvilinear in shape, the trend attempts to fit, and in so doing summarize, the overall pattern exhibited by the full set of values.

BARS

A bar can be thought of as a line with the second dimension of width added to the line's single dimension of length, which transforms the line into a rectangle. The width of the bar doesn't usually mean anything; it simply makes each bar easy to see and compare to the others. The quantitative information represented by a bar is encoded by its length and also by its end in relation to a quantitative scale along an axis. So far, each of the graphical objects that we've examined encodes quantitative values as 2-D position relative to quantitative scales along one or both axes. Tick marks (i.e., the little lines that mark the location of values along the axis, like the little lines on a ruler) and their numeric labels enable us to translate the position of objects such as points, lines, or bars into quantitative values.

Take a moment to look at the bars on the graph below:

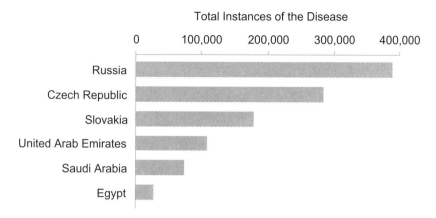

FIGURE 6.6 This graph uses bars to encode quantitative values.

Remember that the width of the bars has no meaning. I emphasize this because we have a tendency, as a result of the nature of visual perception, to assign greater value to things that are bigger in total area; however, nothing about a bar other than its length and the point at which it ends should influence our perception of its value. If the bar that represents the number of instances of the disease in Slovakia were wider than the others, we would be tempted to assign it a greater value than its length alone represents. Keep this in mind when you use bars in graphs, making sure that they are all of equal thickness so that the reader's focus is drawn exclusively to their length for quantitative meaning. For most of us, this is not a concern because software that generates graphs doesn't generally allow us to make some bars wider than others.

You may think it odd that the bars in this example run horizontally rather than vertically. Bars can run in either direction. There are times, however, when there is a clear advantage to using horizontal bars rather than vertical. We'll cover this topic later in Chapter 10, *Component-Level Graph Design*.

Because the end points of bars encode quantitative values, points (e.g., dots) at the same locations as the bars' end points could replace the bars and convey the same meaning. So why use bars at all? Bars do one thing extremely well: because of their visual weight, they stand out so clearly and distinctly from one another that they do a great job of representing individual values discretely. In the previous example, the beginning of each bar at the axis that provides its categorical label (Russia, Czech Republic, etc.) clearly connects it with that categorical item, and that value alone. This makes it easy to focus on individual values and to compare them, especially when they are positioned next to each other.

Because the length of a bar is one of the attributes that encodes its quantitative value, its base should always begin at the value zero. Notice how the lengths of the bars no longer support accurate comparisons of quantitative values in the

Bars that run vertically are sometimes called *columns*. However, because the purpose of horizontal and vertical bars is exactly the same, I use the term bar for both.

graph below on the right, contrary to the accurate comparisons that are supported by the bars in the graph on the left.

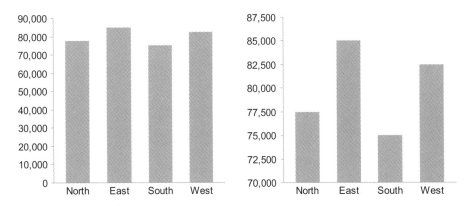

FIGURE 6.7 These graphs illustrate what happens when you use bars to encode data without beginning the quantitative scale at zero.

Comparing the lengths of the bars for the north and east regions in the right-
hand graph suggests that the value in the east is twice that of the north, but this
is far from the case. When we use bars to encode values, the bars themselves
should enable the viewer to make relative comparisons between the values that
they represent, without needing to refer to the quantitative scale.

BOXES

Boxes are the least familiar graphical objects that we'll use to encode values in
graphs. This is because they are exclusively used to compare the distributions of
different sets of values with one another, which is done less often in most
organizations than other quantitative comparisons, except by statisticians and
well-trained data analysts. This is sad because distributions of values have
important stories to tell.

Like a bar, a box is usually rectangular in shape, but, unlike a bar, which
encodes a single value, a box encodes an entire range of values, usually from the
lowest to the highest. Here's a simple example:

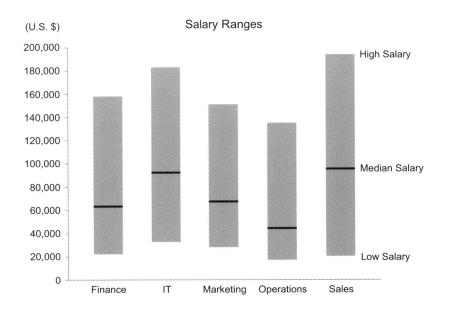

FIGURE 6.8 This graph uses boxes to encode distributions of values.

Box plots are not a standard chart type in Excel. Tricks can be played with Excel, however, to create simple versions of the box plot such as the one to the left and more informative versions as well. For instructions on how to do this, see Appendix E, *Constructing Box Plots in Excel*.

Graphs that use boxes to encode values are called box plots. The mark in the middle (a black line in this case) represents the center of the distribution, usually the median. By focusing on the medians in the previous example, we can quickly see that the typical salary in the Sales Department is higher than in any other, and that the typical salary is lowest in Operations. By focusing on the full lengths of the boxes from bottom to top we can easily see that salaries in the Sales Department are spread across a greater range than in any other. By focusing on the tops and bottoms of the boxes, we can see that the person with the highest salary works in Sales and the person with the lowest salary works in Operations. These simple boxes tell us a great deal about salaries in these departments and how they differ. As such, box plots deserve a place in our repertoire of graphical objects.

SHAPES WITH 2-D AREAS

The only other type of object that is routinely used to encode values is an assortment of 2-D shapes that represent values in proportion to their area (i.e., their two-dimensional size) rather than (or sometimes in addition to) their location in the graph. Here's an example in the form of a graph that has become quite familiar through frequent use:

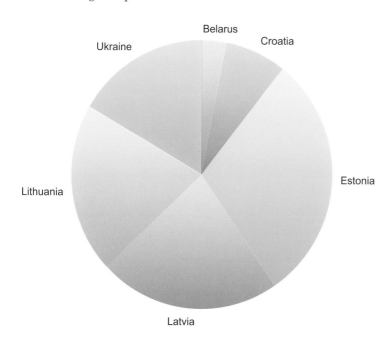

FIGURE 6.9 This graph uses the 2-D areas of slices to encode quantitative values.

As I'm sure you know, this is a *pie chart*. It is part of a larger family of *area graphs*, which encode quantitative values as the sizes of 2-D areas. Pie charts use segments of a circle (i.e., slices of a pie). The size of each piece of the pie is equal to its value compared to the total value of all the slices.

You may be wondering, "where is the axis on this graph?" Good question. Remember, one of the defining characteristics of a graph is that it has at least one axis. Though it certainly isn't apparent, a pie chart does have an axis, but, unlike most graphs, this axis isn't a straight line. In pie charts, the perimeter of the circle serves as a circular axis. Just like any other axis, it could be labeled

In addition to the 2-D areas of slices, pie charts also encode values in two other ways: 1) the angles formed by slices where they radiate out from the center of the pie, and 2) the lengths of the curves along the outer edges of the slices. Unfortunately, none of the three ways that pie charts encode values are easy to decode or compare.

with measures of quantitative value corresponding to tick marks of equal distance all the way around the perimeter, as illustrated below:

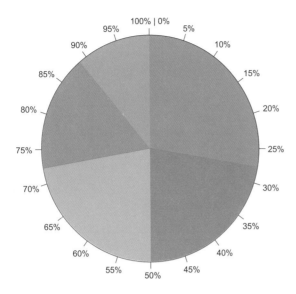

FIGURE 6.10 This is how a pie chart would look if its quantitative scale were visible.

This is never done, however, because it can't be done without making the chart messy and difficult to read.

Speaking of "difficult to read," allow me to make it clear without delay that *I don't use pie charts*, and I strongly recommend that you abandon them as well. My reason is simple: pie charts communicate information poorly. This is a fundamental problem with all types of area graphs but especially with pie charts. Our visual perception is not designed to accurately assign quantitative values to 2-D areas, and we have an even harder time when a third dimension of depth is added. Slices of a pie are particularly difficult to compare but the 2-D areas of even less-challenging geometric shapes, like rectangles, aren't much better. If they're fairly close in size, it's difficult if not impossible to tell which is bigger, and when they're not close in size, the best you can do is determine that one is bigger than the other, but you can't judge by how much. Just for fun, try to put the following slices in order from largest to smallest.

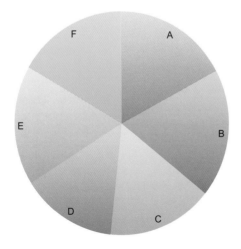

FIGURE 6.11 It is difficult to compare slices of a pie, especially when they're close in size.

You might think that these values are extraordinarily close in size, but as you can see when the same values are displayed in a bar graph, that's not the case.

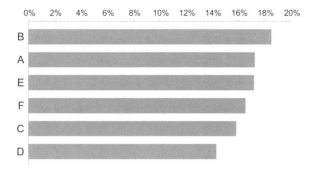

FIGURE 6.12 This bar graph displays the same values that appear in *Figure 6.11* but in a way that makes them easy to compare.

The reason we can see the differences easily in the bar graph but with difficulty in the pie chart is that visual perception is highly tuned for seeing differences among the lengths of objects that share a common baseline but not well tuned for discerning differences among 2-D areas.

Another means of encoding values using 2-D areas involves varying the sizes of points, such as in a scatter plot or on a map. Points that vary in size are usually called *bubbles*. A bubble plot, such as the example below, is simply a scatter plot that uses the vertical and horizontal positions of objects to encode two sets of values, and the sizes of those objects to encode a third set of values.

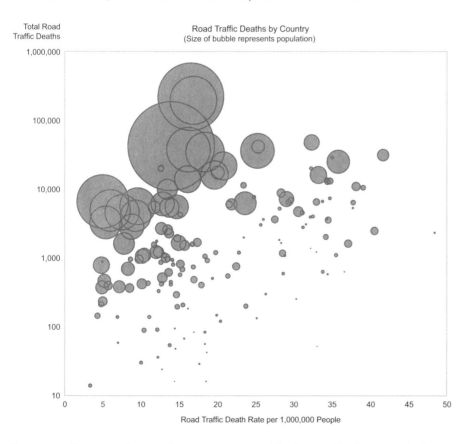

FIGURE 6.13 This is a bubble plot, which uses the size of points (bubbles) to encode a variable.

In a normal scatter plot, you have two axes, and both provide the quantitative values for a single point. Referring back to an earlier example, let's say you want to measure the correlation between employee tenure (years on the job) and employee productivity (e.g., the speed at which one can assemble widgets). In a normal scatter plot, you can display this relationship using the scale along one axis to represent tenure and the scale along the other to measure productivity.

For each employee, you would place a point on the graph at the location where that employee's measures of tenure and productivity intersect. But what if you also need to account for the possible effect of compensation as a third variable? You could try to use a third axis, making the graph 3-D. This is a fairly common method that is supported by many software products, but the simulation of a third dimension on a flat piece of paper or computer screen results in a graph that is extremely difficult to read. We'll look at this further in the later chapters on graph design. Using the 2-D areas of bubbles is an effective way to add a third set of values to a graph but only when approximate values are all that's needed, and the third variable adds enough to the story to justify the additional level of complexity.

VARYING COLOR INTENSITIES

Another method that is sometimes used involves varying intensities of the same color (e.g., light gray to black) to represent values. For example, in the following bubble plot, in addition to horizontal position to represent patients' percentage of medical costs, vertical position to represent per-patient medical costs, and bubble sizes to represent the number of patients, a fourth quantitative variable— patient age groups—is represented by bubble color intensity, from light gray (the youngest patients) to black (the oldest patients).

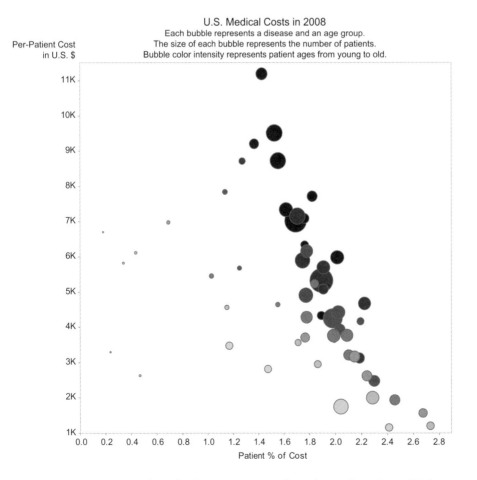

FIGURE 6.14 This bubble plot encodes separate variables using the size and color intensities of the bubbles.

As you can imagine, it takes a little practice to read graphs such as this, which display this many variables simultaneously. When people need to see these variables in relation to one another, however, it is often worth the effort. This particular example tells a rich story. Some of the obvious facts that we can discern are:

- The highest medical costs appear to be associated with older patients, based on the fact that bubbles near the top tend to be dark, and those near the bottom tend to be light.
- Medical treatments for particular age groups and diseases that involve the most patients do not account for the highest per-patient costs, based on the fact that bubbles closest to the top are not the biggest.
- If there is a correlation between per-patient costs in dollars and the patients' percentage of medical costs (as opposed to the portion paid by insurance companies), it appears to be negative. As patients' percentage of costs increases, the total patient costs in dollars tend to decrease.

If we were to dig into this information in greater detail to identify each disease and age group, many more stories would no doubt emerge.

In Chapter 5, *Visual Perception and Graphical Communication*, we considered the reasons why 2-D areas and color intensities don't work as well as the other means of encoding values that we've examined. With rare exceptions, we'll rely solely on the following four objects for encoding values in graphs because they rely on visual attributes that we can easily and accurately perceive: either the lengths of objects or their 2-D positions:

- Points
- Lines
- Bars
- Boxes

Visual Attributes Used to Encode Categorical Items

The same objects that are used to encode quantitative values in graphs (points, lines, bars, and boxes) are also associated with categorical items. In a graph that displays the number of registrations (a measure of quantitative value) by region (a set of categorical items), something about each object that encodes a number of registrations also needs to indicate the region it represents. In the following graph, the position of each bar along the X axis in conjunction with a categorical label identifies the region:

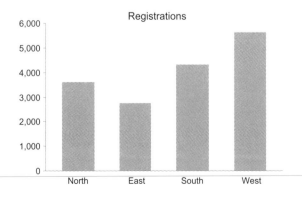

FIGURE 6.15 This bar graph uses position along the X axis to encode categorical items (regions).

Categorical items may be represented in graphs by applying one of the following attributes to the objects that are also used to encode the quantitative values:

- 2-D Position
- Hue
- Point shape
- Fill pattern
- Line style

2-D POSITION

The most common attribute used to identify categorical items is 2-D position. In the following example, there is no question about which month is associated with each of the bars:

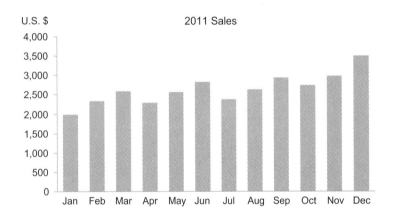

FIGURE 6.16 This graph uses position along a categorical list of months on the X axis to label the bars.

Position is the clearest means to associate quantitative values with categorical items in graphs. Position works equally well when the object that encodes the quantitative values is a point, line, bar, or box.

HUE

The second most effective attribute for associating quantitative values with categorical items is hue. In the next example, because the attribute of position has already been used to identify quarters, we must use a different attribute to add a distinction between the categorical items of Direct versus Indirect sales. Hue will work just fine.

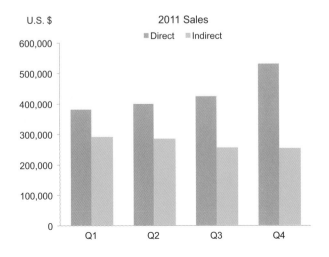

FIGURE 6.17 This graph uses two attributes for encoding categorical items: position along the X axis for the quarters of the year and hue for the distinction between direct and indirect sales.

In this example, a legend has been added to label the two categorical items, Direct and Indirect sales. Interpreting categorical items that are encoded through the use of hue is not quite as direct as interpreting items indicated by position, but once you've read the legend and stored the association of blue with direct sales and orange with indirect sales in your memory, you can easily interpret the graph. In fact, the hues are distinct enough that you can concentrate independently on the blue set of direct sales values or the orange set of indirect sales values with little distraction from the other set. Hue works equally well as an attribute for points, lines, bars, and boxes, as long as the object is not so small that you must strain to distinguish the hues.

POINT SHAPE

This third attribute for encoding categorical subdivisions only applies when points are used to encode the quantitative values. Shapes such as dots, squares, triangles, asterisks, diamonds, and dashes can be used to distinguish quantitative values that belong to different categories. In the graph below, both points and lines are used to encode the quantitative values, with differing symbols for the points to distinguish dollars booked versus dollars billed:

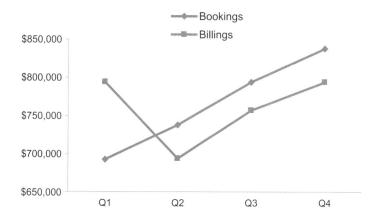

FIGURE 6.18 This graph uses differing symbol shapes to encode categorical subdivisions, in this case diamonds for bookings and squares for billings.

Different point shapes, like the squares and diamonds in the above graph, can be distinguished fairly easily but not as easily and rapidly as different colors. Nevertheless, there are times when using point shapes to encode categorical items comes in handy, such as when you've already used position and color for other categorical distinctions, as in the following graph:

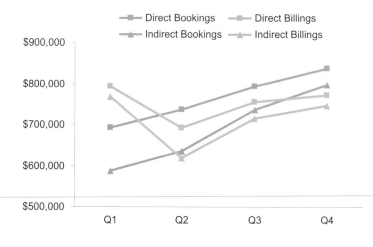

FIGURE 6.19 This graph uses position, color, and symbol shapes to encode categorical items, in this case position for quarters of the year, color for bookings vs. billings, and symbol shapes for direct vs. indirect sales.

FILL PATTERN

Fill patterns are only used to encode categorical items when the quantitative values are encoded as bars or boxes. The 2-D portion of either can be filled with differing patterns to distinguish categorical items, such as Direct and Indirect sales in the following example.

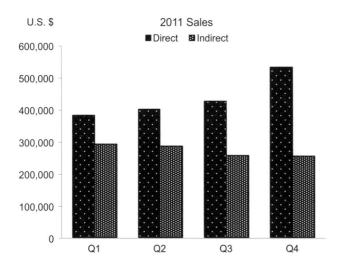

FIGURE 6.20 This graph uses fill patterns, in this case different arrangements of dots, to encode the categorical items of direct and indirect sales.

Only use fill patterns as a last resort. Not only are they harder for our eyes to distinguish than colors, they can also play real havoc with our vision, causing a dizzying effect called *moiré vibration*. This effect is especially strong if the patterns consist of vivid lines running in various directions, as in the next example:

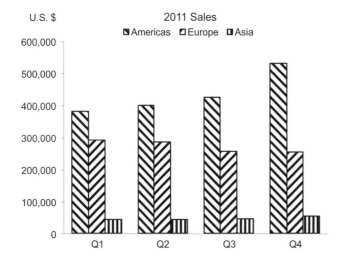

FIGURE 6.21 This graph uses fill patterns that are hard on the eyes and therefore difficult to read.

Try looking at this graph for awhile and you may actually feel ill because of the appearance of vibration created by the lines, so it isn't hard to understand why fill patterns should be avoided whenever possible. If you must use them, do so with great care, muting the patterns and selecting ones that can be displayed together with minimal visual vibration. The primary reason that you may sometimes need to use fill patterns is when you must print your graph on paper or photocopy it for distribution, and color printing is not an option. In such cases, if you only need a few categorical distinctions, distinct shades of gray (e.g., black, dark gray, medium gray, and light gray), which are variations in color intensity, usually work better than fill patterns.

LINE STYLE

Varying line styles can be used to encode categorical items when the quantitative values are encoded as lines. Lines may be solid, dashed, dotted, and so on. Here's an example:

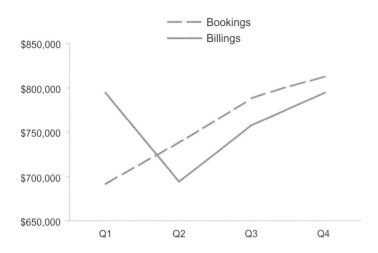

FIGURE 6.22 This graph uses line style, in this case a solid and a dashed line, to encode the categorical items of bookings and billings.

Line styles don't work as well as colors because lines are usually used to represent the pattern in an entire series of values, so breaks in the lines undermine our perception of the continuous series of values as a whole. Consequently, like fill patterns, line styles are best reserved for occasions when color printing is not an option. Differing the color intensity or thickness of lines usually works better.

Relationships in Graphs

Graphs display relationships in quantitative information by giving shape to those relationships. Graphs vary primarily in terms of the types of relationships they are used to communicate, so it is useful to understand the specific types of relationships that graphs can display. With this knowledge, you will be able to quickly match the quantitative message you wish to communicate to the structural design that can do the job most effectively.

Essentially, there are eight types of relationships that we typically use graphs to display:

- Time series
- Ranking
- Part-to-whole
- Deviation
- Distribution
- Correlation
- Geospatial
- Nominal comparison

The categorization of graphs into five of these relationship types was proposed by Gene Zelazny (2001) *Say It with Charts*, Fourth Edition. McGraw-Hill. His taxonomy includes the following: 1) component comparison (i.e., part-to-whole), 2) item comparison (i.e., ranking), 3) time-series comparison, 4) frequency distribution comparison, and 5) correlation comparison. I've expanded this list to include nominal comparison, deviation, and geospatial relationships.

Time Series

A time-series relationship is nothing more than a series of quantitative values that feature how something changed with time, such as by year, quarter, month, week, day, hour, minute, or even second. This relationship deserves to be treated separately because of the unique characteristics of time when displayed graphically. Graphs are a powerful way to view patterns and trends in values over time, which is why they are commonly used to display time series. A typical example is the closing price of a stock on each day of the last three months.

To determine whether your quantitative message involves a time series, first state the message in writing, and then look to see whether you used any of the following words:

- Change
- Rise
- Increase
- Fluctuate
- Grow
- Decline
- Decrease
- Trend

These words, with the exception of "trend," were suggested by Gene Zelazny (2001) *Say It with Charts*, Fourth Edition, McGraw-Hill. The subsequent lists of words that accompany each relationship type below were derived primarily from the same source.

If you did, then your message likely involves a time series.

Ranking

A graph displays a ranking relationships when it displays how a set of quantitative values relate to each other sequentially, sorted by size from lowest to highest or vice versa. A typical example is a ranking of donors based on the amounts of their donations.

Words and phrases that suggest a ranking relationship include:

- Larger than
- Smaller than
- Equal to
- Greater than
- Less than

Part-to-Whole

A graph displays a part-to-whole relationship when it features how individual values that make up the whole of something (e.g., total sales) compare to one another and how each compares to the whole. The individual values (the parts) compare to the whole as ratios, which may be expressed either as percentages that sum to 100% or as rates that sum to 1. A typical example is the percentage of sales attributed to various sales regions, which together add up to total sales.

Words and phrases that suggest a part-to-whole relationship include:

- Rate or rate of total
- Percent or percentage of total
- Share
- Accounts for X percent

If your message uses one of these expressions, you are likely presenting a part-to-whole relationship.

Deviation

A graph displays a deviation relationship when it features how one or more sets of quantitative values differ from a reference set of values. The graph does this by directly expressing the differences between the two sets of values. Common examples include the following:

- The degree to which actual worker productivity differs from target productivity
- The degree to which donations over time differ from donations at some specific time in the past
- The degree to which headcount in each month varied from headcount in the previous month
- The degree to which sales of various products differ from sales of a particular product

Deviations are usually expressed using one of the following units of measure:

- Positive or negative amounts relative to the reference values
- Positive or negative rates or percentages relative to the reference values (i.e., where the reference values always equal 0 or 0%, and all other values are expressed as plus or minus rates or percentages)
- Rates or percentages relative to the reference values (i.e., where the reference values always equals either a rate of 1 or 100%, and all other values are expressed as relative rates or percentages)

Words and phrases that express deviation relationships include:

- Plus or minus
- Variance
- Difference
- Relative to

Distribution

A graph displays a distribution relationship when it features how quantitative values are distributed across an entire range, from the lowest to the highest. When a graph displays the distribution of a single set of values, the relationship is called a *frequency distribution*, for it shows the number of times something occurs (i.e., its frequency) within consecutive intervals over the entire quantitative range. Graphs of this type visually display the same type of information that is communicated by the range and standard deviation measures of variation. For instance, you may want to display the number of employees whose salaries fit into each interval along a series of salary ranges from lowest to highest. It wouldn't make sense to count the number of employees for every distinct salary amount because there are far too many. Rather, you would break the entire range of salaries into smaller ranges (a.k.a. intervals), such as "less than $15,000," "greater than or equal to $15,000 and less than $30,000," "greater than or equal to $30,000 and less than $45,000," etc., and then count the number of employees that fit into each.

Words that suggest a distribution relationship include:

- Frequency
- Distribution
- Range
- Concentration
- Normal curve, normal distribution, or bell curve

Correlation

A graph displays a correlation when it is designed to show whether two paired sets of quantitative values vary in relation to each other, and, if so, in which direction (positive or negative) and to what degree (high or low). These graphs display the same type of information that is expressed by a linear correlation coefficient. A typical example is a correlation of marketing expenditures and subsequent sales. The existence of a correlation may indicate that one thing causes the other (i.e., a causal relationship), but not necessarily. Two variables can also be correlated because they are both caused to behave as they do by one or more external factors.

Words that indicate a correlation include:

- Increases with
- Decreases with
- Changes with
- Varies with
- Caused by
- Affected by
- Follows

Geospatial

In a display that features where quantitative values are located, the relationship is primarily spatial. Most spatial relationships feature the geographical location of values by displaying them on a map. A typical example is a map that displays how many cases of a particular disease occurred in each country throughout the world. Spatial relationships are not limited to geography, however. For example, someone who manages environmental conditions (e.g., air temperature) throughout a large, multi-story building, might want a graphical display that positions values on a floor plan of the building. Because nongeographic spatial displays of quantitative data are rare, however, we'll focus only on geospatial displays.

Words that indicate a geospatial relationship:

- Geography
- Location
- Where
- Region
- Territory
- Country
- State
- City

Nominal Comparison

A nominal comparison relationship is the simplest of all, and also the least interesting, which is why I saved it for last. In this relationship, items along the categorical scale have no particular order because they represent a nominal scale. The goal is merely to display a set of discrete quantitative values so they can be easily read and compared. "This is bigger than that," "this is the biggest of all," "this is almost twice as big as that," and "these two are far bigger than all the others" are some of the messages that stand out in nominal comparison relationships. An example could be the relative amounts of employee turnover for a group of departments in a single month, quarter, or year. You might wonder why we wouldn't sort the values from largest to smallest or vice versa to display a more interesting ranking relationship. That might be useful, but people sometimes become accustomed to seeing a set of categorical items arranged in a particular order even though that order is arbitrary, such as geographical regions always arranged as north, east, south, and west. It might be confusing for your audience to see those regions in any other order, in which case you might stick with a nominal comparison relationship.

Graph Design Solutions

You now know the types of relationships that graphs can be used to display as well as the visual objects and attributes that are available for encoding data in graphs and some of the strengths and weaknesses of each. The next step is to learn the best structural solution for displaying each type of relationship. Let's examine the relationship types and think through the design solutions for each. Approaching the solutions in this manner will give you the required design tools in a way that deepens your understanding of them, enabling you to call them quickly to mind whenever needed.

Before we tackle the first type of graphical relationship, here's a reminder of the different objects that can be used to encode quantitative values:

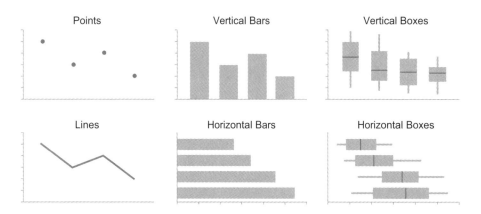

FIGURE 6.23 These objects are commonly used to encode quantitative values in graphs.

Review each for a moment to remind yourself of their individual strengths for displaying quantitative messages.

Nominal Comparison Designs

We'll begin with the simplest quantitative relationship. A nominal comparison graph displays a series of discrete quantitative values to highlight their relative sizes. Because the values are discrete, each relating to a separate categorical subdivision with no connection between them, you want to encode the quantitative values in a way that emphasizes their distinctness. Which visual encoding object emphasizes the distinctness of each value best? Bars do because they are quite distinct from one another, each carrying a great deal of visual weight. Graphs like the following example are quite common:

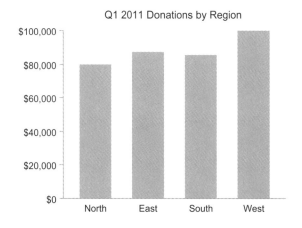

FIGURE 6.24 This is a nominal comparison relationship graph that displays donations by region.

Even though the bars in this example are vertical, horizontal bars would work just as well.

When bar graphs consist entirely of fairly long bars that are similar in length, it can be difficult to discern the subtle quantitative differences between them. Narrowing the quantitative scale so that it begins just below the lowest value and ends just over the highest value could help to make these subtle differences easier to see, but bars require a zero-based scale. So when there are subtle quantitative differences to display, you can usually replace the bars with points (e.g., dots), which can display the values distinctly but don't require a zero-based scale as bars do. Graphs of this type usually go by the name dot plots. The following example is a dot plot of the same data that as shown in Figure 6.24:

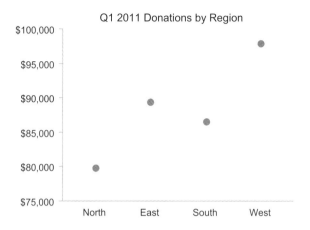

FIGURE 6.25 Dot plots are an effective alternative to bar graphs when you want to narrow the quantitative scale such that it no longer begins at zero.

Nominal comparison relationships can be effectively encoded using any of the following objects:

- Vertical bars
- Horizontal bars
- Points

Time-Series Designs

A time-series graph displays quantitative values along multiple, sequential points in time. Consequently, one axis of the graph provides the time scale, with labels for each interval of time (years, quarters, etc.). Intervals of time have a natural order. You would almost never display time other than chronologically.

When you imagine a visual representation of time, how do you see it arranged on the page? Because of an old convention in most cultures, there is only one way to lay out time on a page that wouldn't seem strange or confusing: horizontally, from left to right along the X axis. Given this fact, for time-series graphs we can eliminate two of the encoding methods shown in *Figure 6.23*. Take a look and determine which they are.

. . . .

Vertical bars and vertical boxes can both be used to display time-series values, but horizontal bars and horizontal boxes should not be used. That's because horizontal bars and horizontal boxes would use the X axis to label the quantitative scale and the Y axis to label intervals of time (e.g., months). Time-series graphs should always use the horizontal axis for the time scale and the vertical axis for the quantitative scale.

We've eliminated two graphical encoding methods, which leaves us with four more from which to choose. Keep in mind as we continue that there may be more than one encoding method that works.

Vertical bars work in time-series graphs but should only be used when you wish to emphasize and compare the values associated with individual points in time rather than the overall pattern of values as they change through time. This is because the height and visual weight of the individual bars distract from the overall shape of the data. To trace the shape of change through time along a series of bars requires that we focus exclusively on the ends of the bars, which is difficult to do.

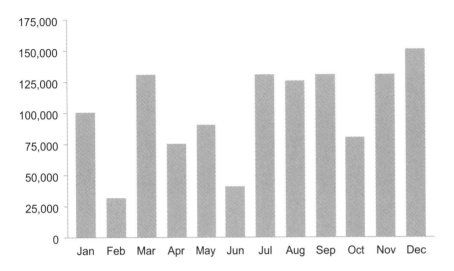

FIGURE 6.26 This graph uses bars to encode time-series values, which is not effective if you wish to clearly show the pattern of change through time.

There is another method that we can eliminate that doesn't lend itself well to a time-series display. Looking at the available objects for encoding data once again, which do you think would not be an efficient means of displaying left-to-right quantitative values related to time?

. . . .

If you imagine a graph that uses points alone to encode quantitative values across time, you will see that points, floating in space, don't help us visualize the sequential nature of time. Here's an example:

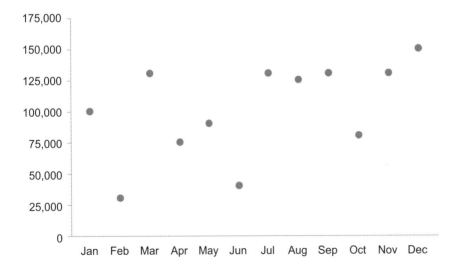

FIGURE 6.27 This graph uses points alone to encode quantitative values, which is not very effective for a time-series graph.

This graph does not give the sense of continuity that is required for displays of time, but adding a line to connect the points corrects this problem.

FIGURE 6.28 This time-series graph uses a combination of points and lines to effectively encode quantitative information.

With this simple change we have made the flow of time visible. Would we still have an effective display of time if we removed the points and used only the line? Absolutely. The points are only useful when it's necessary to show the exact position of values along the line, which is not usually the case. In Chapter 10, *Component-Level Graph Design*, we'll consider occasions when using points or other means to precisely mark the location of values along lines is useful.

There is one occasion when it actually makes sense to display time-series values as points without connecting them with a line. That is when the values in the series do not represent consistent intervals along time. The following graph displays toxin levels that were measured in a particular stream on a few days over a two-month period. By using a line to connect values that were collected at irregular intervals of time, a pattern of change is implied that might be quite different from the pattern that would appear had the values been collected at regular intervals, such as daily.

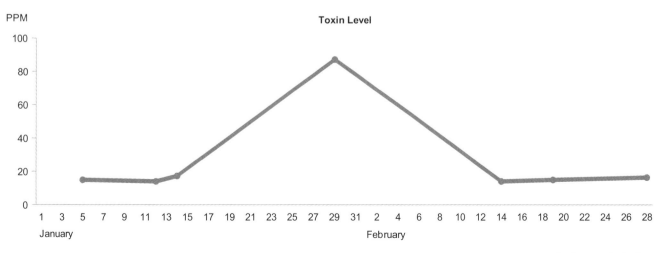

FIGURE 6.29 This line graph displays values that were not collected at consistent intervals of time.

Imagine that toxins were measured each day during this period and the actual pattern of change looks like the following.

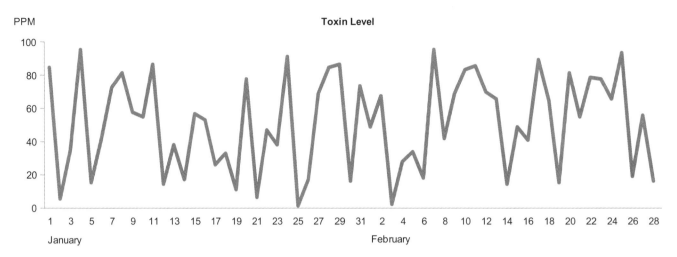

Every point in *Figure 6.29* has a corresponding value in *Figure 6.30*, but the pattern that appears in the first graph based on values that were collected at irregular intervals looks quite different from the pattern based on daily values, which does a much better job of representing what actually happened. When values were collected at irregular intervals of time, it usually works better to omit the line and display only the points in the form of a dot plot, illustrated below.

FIGURE 6.30 This line graph displays values that were collected daily during the same period of time that appears in *Figure 6.29*.

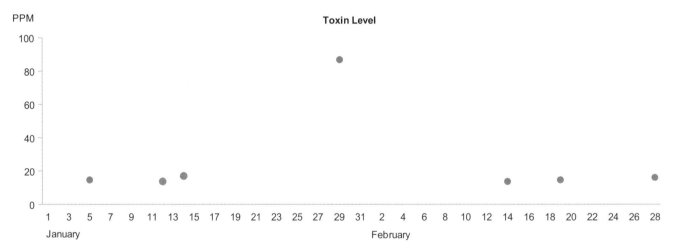

In addition to the line graphs, vertical bar graphs, and dot plots, when a graph shows how the distribution of a set of values changes through time, box plots or individual data points in the form of *strip plots* may also be used. We'll look at box plots and strip plots soon in the *Distribution Designs* section.

Any of the following can be used to encode quantitative values in time-series displays:

FIGURE 6.31 Because the values in this time series were not collected at regular intervals of time, it works better to show the individual values as points only, without connecting them with a line.

- Vertical bars (when you want to emphasize individual values, rather than the pattern of an entire series)
- Lines (with or without points when you want to show the pattern of change through time)

- Points only (when displaying values that were collected at irregular intervals of time)
- Vertical box plots (when displaying how distributions changes through time)

Ranking Designs

A ranking graph displays how discrete quantitative values relate to one another sequentially by magnitude, from low to high or high to low. One axis of the graph must label the categorical items and the other must provide a quantitative scale. Because we want to emphasize each individual value and allow the reader to easily see its rank compared to that of any other value, we should encode the values using a graphical object that visually reinforces the individuality of the values and their relative sizes. Which of the available graphical objects do this best? Bars do.

Here's a graph that uses bars, but so far it is only designed to display a nominal comparison relationship, with the categorical items sorted alphabetically:

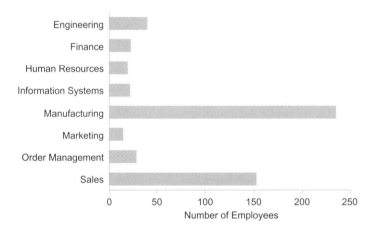

FIGURE 6.32 This graph uses bars to encode the values, which are the appropriate choice for a ranking display, but the bars are not arranged in a manner that highlights the ranking relationship.

Given the usefulness of seeing a ranking relationship based on the number of employees associated with each department, this graph works much better if we arrange the departments as follows.

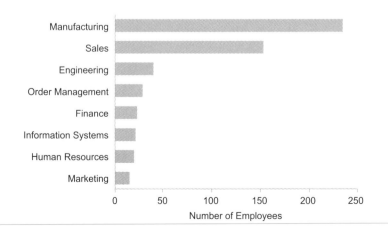

FIGURE 6.33 This ranking relationship is designed very effectively as a bar graph with the bars sorted by size.

Now the ranking message is crystal clear.

Can you think of any reason why vertical bars, as opposed to the horizontal bars, would not work as well? Both types of bars work, but there are times when horizontal bars may be preferable to vertical bars, and vice versa. We'll take a look at this later in Chapter 10, *Component-Level Graph Design*. For now, though, the following rules of thumb will come in handy:

Purpose	Sort Order	Bar Position
Highlight the highest values	Descending	Vertical bars: highest bar on left Horizontal bars: highest value on top
Highlight the lowest values	Ascending	Vertical bars: lowest bar on left Horizontal bars: lowest bar on top

In most cultures, we tend to think of the top, as opposed to the bottom, and the left, as opposed to the right, as the beginning. This convention might be rooted in written languages that are read across the page from left to right and top to bottom, or might actually be wired fundamentally into visual perception, as has been suggested by some recent research. Until further research is done, the jury is still out on this matter.

Just as in graphs showing nominal comparison relationships, in a ranking relationship graph, bars can be replaced with points when it is useful to narrow the quantitative scale and in so doing to remove zero from the graph's base. The following graphical objects can therefore be used to encode quantitative values in ranking graphs:

- Bars (vertical or horizontal, except when the quantitative scale does not begin at zero)
- Points (especially when the quantitative scale does not begin at zero)

Part-to-Whole Designs

As the name suggests, part-to-whole graphs relate parts of something to the whole. They display the proportion of the whole that each part contributes. The best unit of measure for expressing proportions is usually percentage, with the whole equaling 100% and each part equaling a lesser percentage corresponding to its value relative to the whole. In common practice, pie charts are usually used to display part-to-whole relationships, but I've already explained my objections to the use of pie charts and other area graphs in the earlier section *Shapes with 2-D Areas*. People also often use *stacked bar graphs* to display part-to-whole relationships, but a single stacked bar does the job only slightly better than a pie chart. Here are two examples of a stacked bar graph, one vertical and one horizontal:

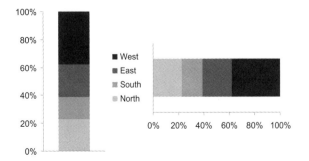

FIGURE 6.34 These examples show the same part-to-whole relationships expressed using two versions of a stacked bar graph.

Rather than comparing these two graphs to determine which works best, take a moment to look for characteristics that hinder their effectiveness.

· · · · · · ·

Did you notice that it is easy to determine the percentage value associated with the north region but not as easy to determine the value of any other region? The value of the north region is easy because all you need to do is look at the percentage scale on the axis and the value is right there. However, for the east region you must look at the values associated with the beginning and end of its portion of the bar (approximately 40% and 62%), then subtract the smaller from the larger value to get its percentage (approximately 22%). In other words, you have to do some math to determine the value. Now see whether you can tell which has the larger percentage, the north or the east region. It's difficult to tell, because they appear to be about the same size. In fact, the east region is a little larger. It's hard to see this because of the way the north and east regions are stacked in the bar.

These problems can be solved by unstacking the bars. Look at these new graphs of exactly the same information, this time displayed using individual bars:

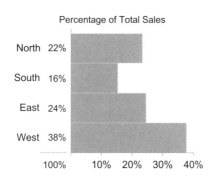

FIGURE 6.35 These examples show the same part-to-whole relationship expressed using individual bars to clearly display the contribution of each part to the whole.

This is better, isn't it? It is easy to interpret the percentage value of each region, and it is easier to tell that the east region has a slightly larger value than the north. It would be nice, however, if we could make it even easier to compare values that are close in size. What could we change about these graphs to accomplish this? We could sort the regions in order of size, thereby placing those of similar size next to each other, which should make the differences easier for the eye to detect. Let's try it:

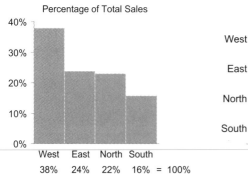

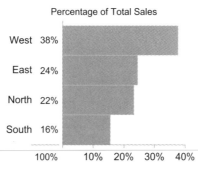

FIGURE 6.36 These two examples display the same part-to-whole information that was displayed in *Figure 6.35*. This time the bars, each representing a different sales region, are sorted by value to make it easier to detect their relative sizes.

Now it's easy to compare the regions to one another. By sorting the regions by size, we've combined a display of ranking and part-to-whole relationships into a single graph.

The one thing that a pie chart has going for it that a bar graph lacks when used for a proportional relationship is the fact that people immediately recognize a pie chart as a part-to-whole display. Unlike a pie chart, which is used exclusively for displaying parts of a whole, bar graphs are used for several purposes. For this reason, when you use a bar graph to display parts of a whole, it is helpful to do a little extra to make this obvious. At a minimum, we should always give the graph a title that states that the bars represent parts of some whole. In the examples above, the title clearly states that the values add up to total sales (i.e., the whole of sales). Something else that I've found helpful is to cause the bars to touch. Because bars usually have spaces between them, people notice when they don't. This clues people into the fact that there's something different about the graph. The fact that the bars touch visually indicates that they are connected to one another, which is indeed the case because they are parts of the same whole. I also often include the precise value of each bar as text, below the labels in vertical bar graphs and to the right of the labels in horizontal bar graphs, not because precise values are needed to read the graph, but because I want to include "100%" as the total, which makes the part-to-whole nature of the graph crystal clear.

Despite the disadvantages associated with stacked bars, one type of quantitative message justifies their use: when you wish to display the whole using a unit of measure other than percentage (e.g., U.S. dollars) but also wish to provide an approximate sense of the relative proportions of the parts. In the next example, the primary message is total sales per channel in U.S. dollars, but the use of stacked bars conveys the secondary message of relative regional sales as well.

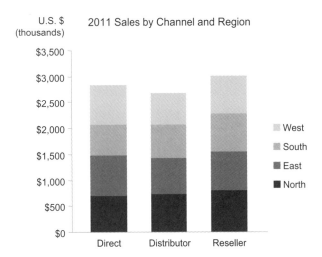

FIGURE 6.37 This is an example of a circumstance when stacked bars are useful.

Even though the stacked bars can't present the parts ideally, they work well enough here to provide additional useful information.

Our investigation leaves us with two variations of a single graphical object for encoding quantitative values in part-to-whole graphs:

- Bars (vertical or horizontal)
- Stacked bars (vertical or horizontal, when you want to feature totals and provide an approximate sense of their parts)

Deviation Designs

A deviation graph displays the degree to which one or more sets of quantitative values differ in relation to a primary set of values. Imagine that you want to provide a way for the Director of Human Resources to see how the number of employees in each department (headcount) differs from the plan. You could display this information in the following way:

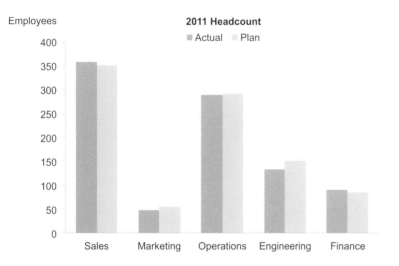

FIGURE 6.38 This graph compares actual headcount to the plan for each department.

However, if the director is only interested in how much actual headcount varies from the plan, this graph forces her to work harder than necessary, doing calculations in her head, to discern the differences. Why not display the differences directly? The next graph does just that.

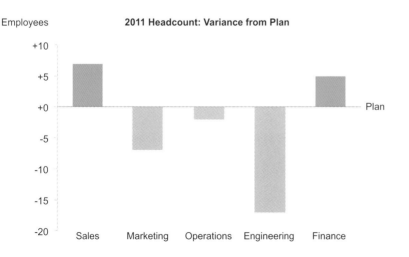

FIGURE 6.39 This graph directly displays the variance of actual headcount from the plan.

A graph should always display as directly as possible the information that people need. In the preceding example, positive headcount variances, which are usually what organizations try to avoid most, are highlighted in red to draw attention to them as unfavorable. Sometimes people want to see deviations expressed in raw numbers, such as the actual number of employees shown in the previous graph, and sometimes they want to see them as percentages, as in the example below.

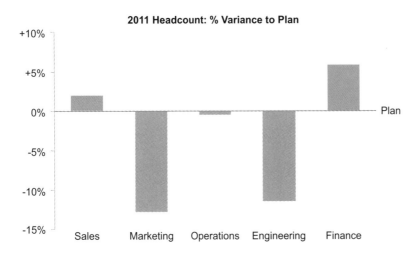

FIGURE 6.40 This graph directly displays the variance of actual headcount from the plan as a percentage.

Both graphs above are valid, but they tell different stories. In the case of *Figure 6.39*, viewing the actual number of employees above and below the headcount plan would help the director understand the overall impact of the variance on the organization while the percentage variance would help her compare how well each of the departments is performing in relation to the plan.

Deviation relationships are often teamed with other relationships, especially time series. Displaying the difference between various measures and a reference measure over time is common practice in the workplace. When combined with a time-series relationship, deviations are often encoded as lines to represent the continuous, flowing nature of time. When combined with other types of relationships (e.g., ranking and part-to-whole), or when functioning on their own, deviation relationships usually use bars. The following example combines a deviation and a time-series relationship. In this case, a line is used to show how the variance changes through time.

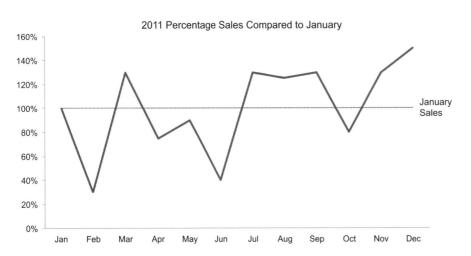

FIGURE 6.41 This combination of a deviation and a time-series relationship displays the variance of monthly sales from January sales, expressed as percentages.

If you wanted to display the relationship of each month's sales to those of the prior month, the graph would look something like this:

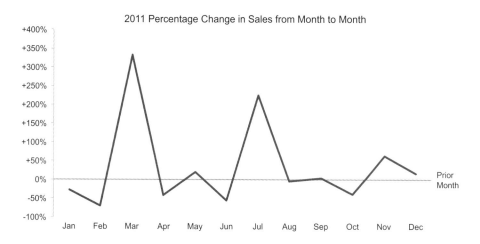

FIGURE 6.42 This combination of a deviation and a time-series relationship displays the variance between sales in a given month and those of the previous month, expressed as positive and negative percentages.

Our investigation has shown that deviation relationships can be effectively displayed using the following objects:

- Bars (vertical or horizontal)
- Lines (when displaying deviation through time)

Distribution Designs

A distribution graph displays the way in which one or more sets of quantitative values are distributed across their full quantitative range from the lowest to the highest and everything in between. The best type of graph to use depends in part on whether we need to display the distribution of a single set of values or distributions of multiple sets so that they can be compared.

SINGLE DISTRIBUTION

A graph of a single distribution shows how often something occurs, expressed either as a count or percentage, distributed across a series of consecutive, quantitative ranges (a.k.a. intervals). The type of graph that works best for displaying a single distribution varies somewhat, depending on whether you want to emphasize the number of occurrences in each interval of the thing being measured or the overall shape of the distribution across the entire range. Given what you know about the unique strengths and weaknesses of each available object for encoding quantitative values, take a moment to determine which would work best for each of the following quantitative messages:

1. The number of orders that fall into each of the following dollar ranges:
   - Less than $5,000
   - Greater than or equal to $5,000 and less than $10,000
   - Greater than or equal to $10,000 and less than $15,000
   - Greater than or equal to $15,000 and less than $20,000
   - Greater than or equal to $20,000 and less than $25,000

2. The time it is taking overall to ship orders from the warehouse once they've been received, based on the following intervals:

- 1 day
- 2 days
- 3 days
- 4 days
- 5 days
- 6 days
- 7 days
- 8 days

. . . .

Given the emphasis in the first message on the frequency of occurrences in each of the dollar categories, which graphical object would work best to highlight these individual measures? Both bars and points highlight individual values, but bars, because of their greater visual weight, do a better job of making individual values stand out. A graph that uses bars to display a distribution is called a *histogram*. Here's an example of the first scenario above displayed as a histogram:

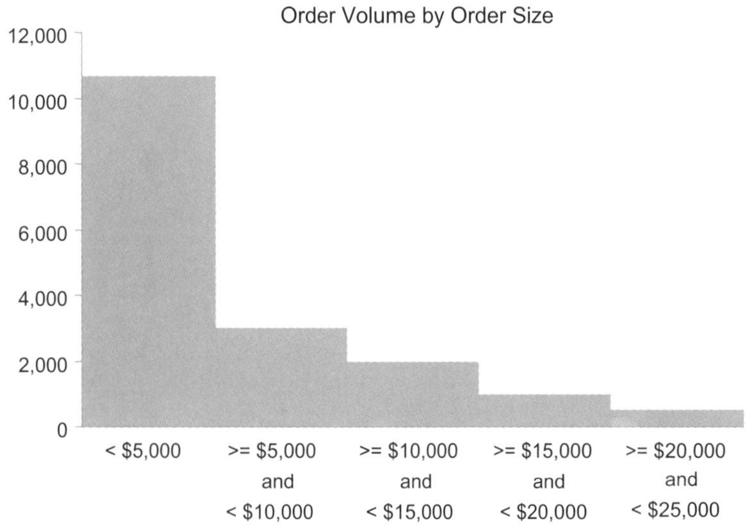

FIGURE 6.43 This histogram displays a single distribution. The bars in a histogram do not conventionally have spaces between them, which visually suggests the continuous rather than discrete nature of the scale.

Histograms are by far the most common way to display distributions. Although horizontal bars work just as effectively, vertical bars are used more often and are therefore more familiar.

Given the emphasis in the second scenario above on the shape of the distribution rather than on values of particular intervals, which graphical object would most effectively display this information? Lines draw the shape in simple visual terms. The name for a graph that uses a line to encode the shape of a frequency distribution is a *frequency polygon*. Here's an illustration of the second scenario, displayed as a frequency polygon:

FIGURE 6.44 This graph, known as a frequency polygon, uses a line to emphasize the shape of a distribution.

Even though the shape of the distribution can be seen in a histogram as well, a frequency polygon focuses the viewer's attention exclusively on the shape by eliminating any visual component that would draw the eye to the values of the individual intervals. The slight disadvantage of frequency polygons is the fact that they cannot be used as easily as histograms to compare the number or percentage of values in one interval to another. Because histograms do an adequate job of showing the shape of a distribution and excel for comparing the number or percentage of values in one interval to another, they are preferred for general use.

In Chapter 2, *Simple Statistics to Get You Started*, I mentioned a name for a distribution that has a shape similar to the one formed by the line in this graph; it is called a bell-shaped or normal curve.

A frequency polygon works superbly for cumulative distributions. It's sometimes useful to show cumulative occurrences from interval to interval along the entire distribution starting from one end and continuing through the other. For example, in addition to seeing shipping performance in *Figure 6.44*, it might be useful to see the total percentage of orders that shipped in one day, two days, three days, and so on up to eight days when all order have shipped, which would look like this:

FIGURE 6.45 This frequency polygon is based on the same data as *Figure 6.44* but displays cumulative percentages for shipping periods from one through eight days.

A cumulative distribution sometimes tells the story in a way that directly addresses your audience's interests.

Another type of graph, which uses points to display a single distribution, is called a *strip plot*. Unlike histograms and frequency polygons, strip plots display each value in the data set rather than aggregating the number or percentage of values into intervals. The following example displays 25 people in an office by age:

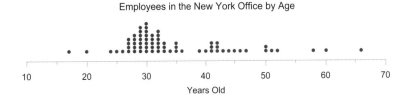

FIGURE 6.46 This graph, called a strip plot, uses individual points for each value in a data set.

Strip plots don't show the shape of a distribution very well, but they're especially useful when you have a small set of values and you want show precisely where each value falls along the quantitative scale. Looking at the example above, you might notice that if 10 employees were 30 years old, you would only see a single point, because all 10 would be located in the same exact space. The best way to solve this problem when you have relatively few values is to *jitter* them. Jittering is the act of repositioning points that are on top of one another either horizontally or vertically so that they're no longer on top of one another. For example, points could be jittered vertically to display a larger set of employees by age as follows:

In Excel, a scatter plot can be used to create a strip plot like the one in *Figure 6.46*. To create a strip plot this way, you must associate the quantitative values in the data set with the X axis and a single value such as 1 with the Y axis, which will line the values up in a single horizontal row. The graph can then be cleaned up by removing the Y axis.

Employees in the New York Office by Age

FIGURE 6.47 Points that represent the same ages have been jittered vertically in this strip plot so that each value can be seen.

For large data sets, another way to solve the problem of multiple points in the same location is to make the points transparent. The result is that areas where more values appear in the same location are denser in color than elsewhere, which allows variation in the number of values to be seen. The following example contains profits and losses associated with more than 8,000 orders ranging roughly from a loss of $16,000 to a profit of $27,000.

We can see that most of the values fall roughly within the range of -$2,500 to +$6,000, with a lesser but still large number of values in the range from +$6,000 to nearly +$10,000. Although this strip plot doesn't display differences in the number of values along the scale as precisely as a histogram or frequency

FIGURE 6.48 In this strip plot, transparency has been used so that color intensity reveals variation in the number of points along the scale.

polygon would, it shows us a number of abnormally low and abnormally high values as individual points extending below and above the range where most of the values fall. In a simple strip plot such as this, you can provide more information by adding marks to show the location of the distribution's center (usually the median) as well as other points of interest, such as the 25th and 75th percentiles to show the range in which the middle half of the values fall. A histogram and a strip plot of this distribution could both be shown to provide a richer picture than either alone could display.

We've learned that the following graphical objects can be used to display a single distribution:

- Bars (vertical or horizontal, in the form of a histogram)
- Lines (especially when you want to feature the shape of the distribution, in the form of a frequency polygon)
- Points (especially for small data sets when you want to show each value, in the form of a strip plot)

MULTIPLE DISTRIBUTIONS

It is often useful to display the distributions of multiple data sets in a single graph so that they can be compared. One simple way to do this, which extends a graph that we've already examined, the frequency polygon, is to combine multiple lines into a single graph, with one line for each data set. Here's an example:

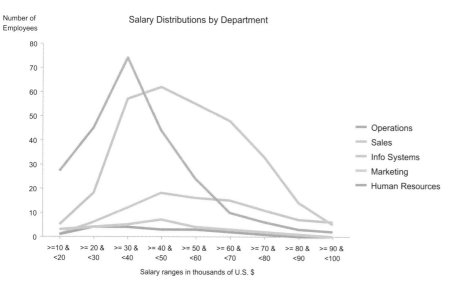

FIGURE 6.49 Lines can sometimes be used to display the distributions of multiple data sets in a single graph.

Rather than counting the number of employees whose salaries fall into each interval along the range, these distributions could also be expressed as the percentage of employees that fall into each. Because the number of employees in each department varies so dramatically, some of the lines in the graph above are relatively flat near the bottom of the scale, which makes it more difficult to see their shapes. If we express the percentage of employees that fall into each interval with the values along each line summing to 100%, the shapes of all distributions are easier to see and compare, as in the example on the next page.

FIGURE 6.50 When the number of values in different distributions varies significantly, it can be useful to use percentages rather than a count of values as the quantitative scale.

Frequency polygons work well for comparing distributions—especially their shapes—as long as there are relatively few lines. Too many lines would look cluttered and would be difficult to read.

When you need to display more than a few distributions or to display how a distribution changes over time, frequency polygons are rarely a good solution. An alternative that works elegantly was invented by John Tukey, a Princeton University statistician who contributed a great deal to the visual presentation of quantitative data, especially to support the exploration and analysis of data. Tukey's solution is called the box plot or *box-and-whisker plot*. What he calls a box is really just a bar—actually a range bar that encodes a range (or distribution) of values from one end of the bar to the other. To this he added a point in the middle (in this case, a horizontal line that divides the box in two) to mark the center of the distribution (usually the median) and two lines, called whiskers: one extending upwards from the top of the box and one extending downwards from the bottom to encode additional information about the distribution's shape. Here's an example of a box and whisker display, with its components labeled:

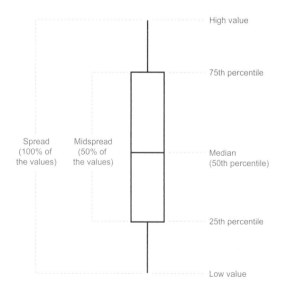

FIGURE 6.51 This is a typical box-and-whisker representation of a single distribution.

As you can see, this simple combination of geometrical shapes communicates quite a lot about a distribution. Here's a list of the full set of facts that it presents:

- The highest value
- The lowest value
- The range of the values from the highest to the lowest (the spread)
- The median of the distribution
- The range of the middle 50% of the values (the midspread)
- The value at or above which the highest 25% of the values reside (the 75th percentile)
- The value at or below which the lowest 25% of the values reside (the 25th percentile)

Now let's look at the sample distribution displayed as a box plot below, to see what we can learn from it.

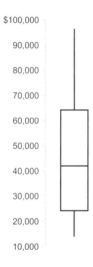

FIGURE 6.52 This example of a salary distribution gives us a chance to practice making sense of box plots.

Assuming that *Figure 6.52* represents a distribution of salaries, the first thing it tells us is that the full range of salaries is quite large, extending from around $14,000 on the low end to around $97,000 on the high end. We can also see that more people earn salaries toward the lower rather than the higher end of the range. This is revealed by the fact that the median, encoded as the short horizontal line in the middle of the box at approximately $42,000, is closer to the bottom of the range than the top. Half of the employees earn between $25,000 and $65,000, which definitely indicates that this distribution is skewed toward the top end of the range. The 25% of employees who earn the lowest salaries are grouped closely together across a relatively small $10,000 range of salaries. Notice the large spread represented by the top 25% of the salaries. This tells us that as we proceed up the salary scale there appear to be fewer and fewer people within each interval along the scale, such as from over $60,000 to $70,000, from over $70,000 to $80,000, and from over $90,000 to $100,000. In other words, salaries are not evenly spread across the entire range; they are tightly grouped near the lower end and spread more sparsely toward the upper end where the salaries are more extreme compared to the norm.

When you need to display multiple distributions for comparison, box plots are hard to beat. Take a minute to study the example below, and see what you can learn about how male and female salaries compare.

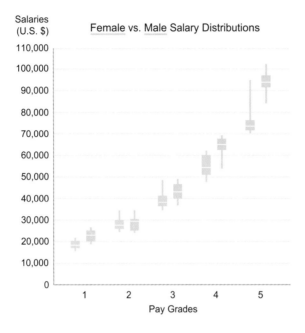

FIGURE 6.53 This box plot separately displays male and female salaries in five different pay grades.

Here are a few of the insights that are revealed in this display:

- On average, women are paid less than men in all salary grades.
- The disparity in salaries between men and women becomes increasingly greater as one's salary increases.
- Salaries vary the most for women in the higher salary grades.

These insights were submitted by Christopher Hanes in response to a data visualization competition that I judged for *DM Review* magazine, which used this graph.

In a new version of the same box plot below, I've made a slight change. Do you see what's different?

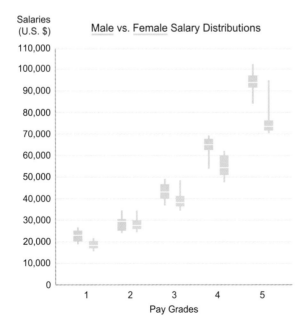

FIGURE 6.54 This box plot is slightly different from the one in *Figure 6.53*, and the difference influences what's emphasized in its story.

All that's different is the order of the male and female boxes. Notice how this simple change alters your perception of the same data. To your eyes, which version of the box plot makes the discrepancy between male and female salaries more apparent? To most people's eyes, the second version highlights the discrepancy more clearly. Why? If you imagine a line being drawn through the medians of all the boxes from left to right using the first version, the line forms a smooth curve. If you do the same with the second version, however, the line is jagged, and that jaggedness is jarring, which increases the salience of the discrepancy. Both of these box plots tell the truth, but one brings this particular aspect of the truth to our attention more.

Before moving on, it's worthwhile to mention that box plots must be used with care for presenting information to others because most people in the world have never learned how to read them. Even if you had never used them before now, you can see that they are easy to understand after a brief introduction, but you won't always have an opportunity to take people through an introduction like the one that I've provided here. For this reason, it is often useful to make box plots as simple as possible. While a five-value box plot (low, 25th percentile, median, 75th percentile, high) might require five minutes of instruction to understand, a simpler three-value box plot requires a few words only. Here's an example of the same salary distribution information as before, this time displayed as simpler three-value boxes:

FIGURE 6.55 This is a simple three-value box plot.

With the simpler version of the graph, the legend on the right might be all that's needed to clarify its use even for people who are not already familiar with box plots. For some, the term "median" might not be clear, so a bit more explanation might be needed, but not much. Obviously, some of the information is lost in reducing the graph from a five-value to a three-value representation of distribution, but for many purposes, the simpler version would suffice.

Box plots are not a standard chart type in Excel, but simple versions can be created by using features that reside in Excel for other purposes. For instructions on creating box plots in Excel, see Appendix E, *Constructing Box Plots in Excel.*

Although boxes are more often oriented vertically, there is no reason why they cannot be arranged horizontally, as illustrated below. If for any reason a horizontal arrangement offers advantages, don't hesitate to use it.

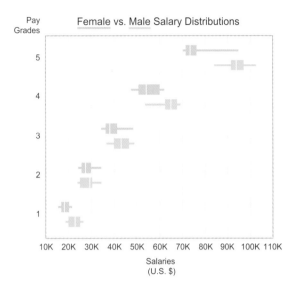

FIGURE 6.56 Boxes may be arranged horizontally or vertically.

Strip plots can also be used to display and compare multiple distributions. In strip plots, multiple distributions are displayed as several rows or columns rather than as a single row or column of points. Strip plots are especially useful when there are relatively few values in the data sets and you have a reason for displaying each value individually. Here's an example, which displays the distribution of student grade point averages (GPAs) for eight schools, one distribution per year for four years.

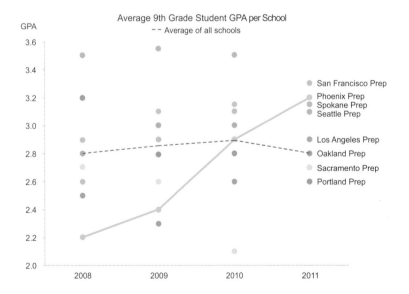

FIGURE 6.57 This strip plot displays the distribution of average GPAs for eight schools by year.

The strip plot displays each school as a separate dot, which makes it possible not only to see the full spread of values in each year from low to high along with a measure of average (the dashed line), but also to see how individual schools performed relative to the others and how the performance of each changed from year to year. In this example, by connecting the four dots that represent Phoenix

Prep's scores with a line, the graph features this school's performance. This view of the data has been customized for administrators at Phoenix Prep, to highlight what's of interest to them. Because this example includes time-series data, it was important to arrange the dots for each distribution vertically so that time could be on the X axis. When time isn't involved, the dots may be arranged either vertically or horizontally.

To summarize, we've learned that multiple distributions can be displayed using any of the following objects:

- Lines (in the form of frequency polygons)
- Boxes (vertical or horizontal)
- Points (arranged vertically or horizontally, in the form of strip plots)

Correlation Designs

Correlation graphs display the relationship between two paired sets of quantitative values to demonstrate whether or not they are related, and, if so, the direction of the relationship (positive or negative) and the strength of the relationship (strong or weak). Because the relationship is between two sets of quantitative values rather than between categorical items and quantitative values, both the X and Y axes of the graph provide quantitative scales. The graph that was created specifically for this purpose is a scatter plot. It is a lot like a strip plot, except that each point is positioned in relation to two quantitative scales—one horizontal along the X axis and one vertical along the Y axis—rather than a single scale. In a sense, just as a strip plot displays a single frequency distribution with one point per value, a scatter plot displays two frequency distributions: one based on the horizontal positions of the points and the other based on the vertical positions.

Let's say that you need to display the potential correlation between the heights of male employees and their salaries. To encode the values for an employee who is 70 inches tall and earns a salary of $60,000, you would find 70 on the X axis, and then move up until you are in line with $60,000 on the Y axis, and then mark that spot on the graph with a point, as illustrated below:

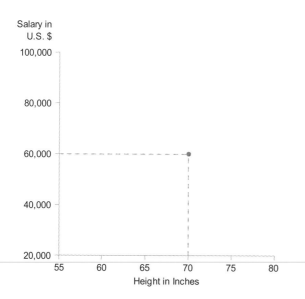

FIGURE 6.58 This illustrates the method used to position correlated values on a graph with X and Y axes.

Based on this illustration, it doesn't take much imagination to recognize that points work perfectly for encoding correlation values. Here's a graph that displays an entire series of paired employee heights and salaries:

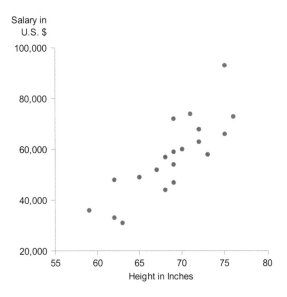

FIGURE 6.59 This graph displays the correlation between employees' heights and their salaries using points to represent the correlated values.

Does the scatter plot above seem to indicate that there is a correlation between these employees' heights and salaries? What could we do to make the potential correlation more visible? We could add a trend line.

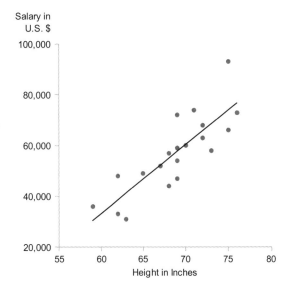

FIGURE 6.60 This scatter plot includes a trend line to highlight the overall pattern of correlation.

It is now easier to see by the upward direction of the trend line from left to right that there is a positive correlation between employee height and salary, but it is not strong; the points are loosely grouped around the line.

Scatter plots are very effective for displaying correlations, with points to mark the values and lines to highlight the pattern. We could end our examination of structural solutions to the display of correlations right here, but a fundamental problem crops up occasionally that motivates us to consider additional solutions.

Think about the audience for the graphs that you prepare. Do they all know how to interpret scatter plots, and, if not, are there those who are not willing to learn? Because it might not be possible to rely on scatter plots in all circumstances, let's look at additional solutions.

Can you come up with a different way to display correlations? Use the example that we've already been working with, the correlation of employee heights and salaries. Can you somehow combine the components available for use in graphs to display the correlation in a way that can be understood more intuitively than a scatter plot?

· · · · · · ·

Were you able to come up with a viable solution? Even if you couldn't, I'm sure the attempt was well worth the effort as a means to reinforce what you've learned about graphs. I'm going to offer a solution that uses two sets of bars to encode the two sets of paired quantitative values. Because we have two quantitative scales, inches for height and U.S. dollars for salary, we need two axes to display them, but we can't use our normal arrangement of an X and Y axis because it wouldn't work to display one set of bars going horizontally and one going vertically. I want to position the bar that represents a particular employee's height and the bar that represents that same employee's salary in a way that clearly associates them with one another. This can be done in more than one way, but the solution that usually works best looks like this:

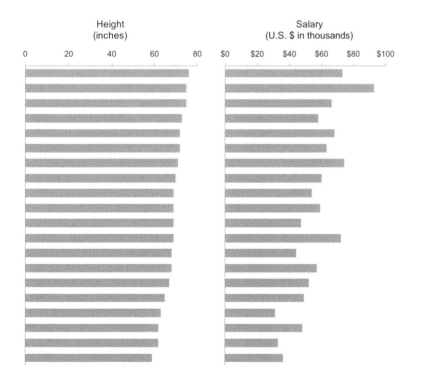

FIGURE 6.61 Two sets of bars, one for each variable, arranged in this manner make it possible to see correlations.

For instructions in how to create a display such as this in Excel, see Appendix D, *Constructing Table Lens Displays in Excel*.

The table lens was first invented by Ramana Rao and Stuart Card. It was designed to do a lot more than the simple examples that I've shown.

Separate sets of bars arranged in this manner for the purpose of revealing correlations are called a *table lens*. In essence, two bar graphs are placed side by side: one displays employees' heights from tallest to shortest and the other displays employees' salaries in the same order. A single row of bars from left to right contains a single employee's height and salary. To look for a potential correlation, you should simply see whether there is a tendency for the salary bars to either decrease or increase fairly consistently as the heights decrease. If the two sets of bars tend to increase together, there is a positive correlation; if salaries tend to increase as heights decrease, there is a negative correlation. Does a correlation appear to exist? Although it isn't perfect, there definitely appears to be a correlation: as height decreases, salary tends to decrease in a corresponding manner. In a table lens, it isn't necessary to use sets of horizontal bars arranged side by side. Sets of vertical bars work as well when you stack them one above the other.

We were able to reveal the same basic information about the correlation between employee heights and salaries in the above graph using horizontal bars as we did in the earlier scatter plot. Scatter plots are superior overall, but if you suspect that your readers will struggle trying to understand a scatter plot, a table lens may produce better results. In addition to being easy to understand, a table lens has another advantage: you can look for correlations among more than two quantitative variables at a time. For example, imagine five sets of bars, one for each variable, rather than two. If your data sets are huge, however, you may be forced to use a scatter plot because you can get only so many bars in a limited amount of space, but thousands of points can fit into a scatter plot.

To summarize, we've learned that correlations can be displayed using the following objects:

- Points (in the form of a scatter plot)
- Bars (in the form of a table lens)

Geospatial Designs

Geospatial displays feature the geographical location of values, positioning values on a map. On geospatial displays, the two best means of encoding quantitative values in graphs—2-D position and the length of objects that share a common baseline—are not available. Because 2-D position is used on a map to represent geographical location, we can't use it to represent a value as we do in scatter plots, dot plots, and line graphs. Because values must be positioned on a map to mark their geographical location, the length of objects such as bars cannot be aligned to share a common baseline. For this reason, we must rely on size, color intensity, and width to encode quantitative values even though these attributes cannot be perceived as precisely as others. Specifically, we'll use the following to encode quantitative values in geospatial displays:

- Points of varying size
- Points or areas of varying color intensity
- Color intensities applied directly to geographical regions
- Lines of varying thicknesses or color intensity

Points can be used to place values in precise locations on a map. The point shape that is typically used on maps is a circle. The quantitative values of points can be encoded by size, color intensity, or both. When you want to encode a single quantitative variable with points on a map, and both size and color intensity are available, size is usually the better choice. It is slightly easier to perceive differences in the sizes of points than differences in color intensities, in part because our ability to perceive color differences decreases with the size of objects, and points are usually fairly small. The number of 911 emergency calls in a particular section of Seattle, Washington have been encoded on the following map by the size of each circle:

911 Emergency Calls in a Section of Seattle

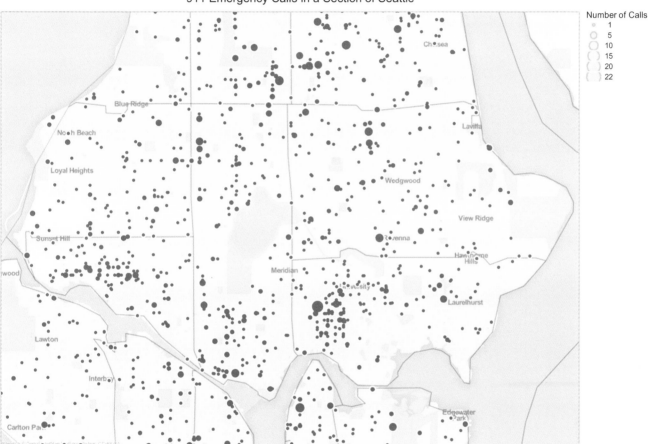

In a case such as this, when the precise location of values must be pinpointed on the map, points are the only means of encoding the values. Because the circles must be relatively small, varying their sizes rather than varying the color intensity of consistently sized circles is the best solution.

Points can be used not only to encode values at specific locations but also to encode aggregated values for entire regions. In the example on the following page, sales revenues are shown at the state level.

FIGURE 6.62 Points that vary in size work well for displaying values at specific locations on a map. This geospatial visualization and those that follow were all created using Tableau.

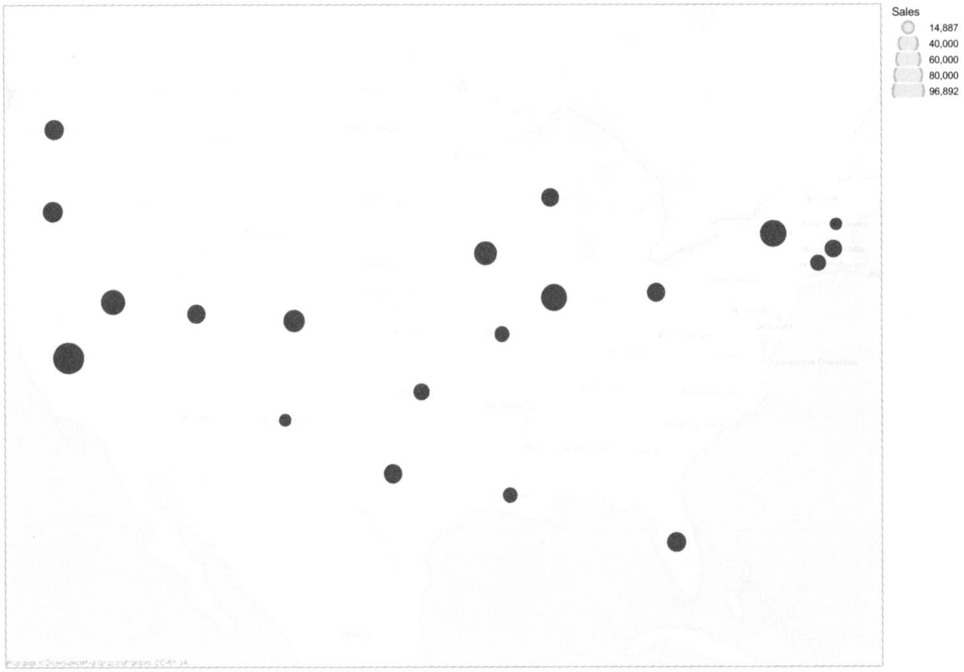

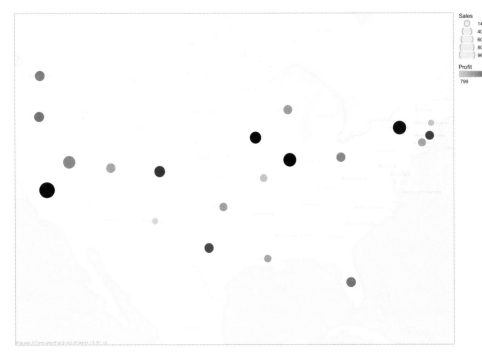

FIGURE 6.63 Points that vary in size can be used to display values for entire regions, such as states.

Because the circles can vary by size and color intensity at the same time, each attribute can be used to encode a separate set of values. In the example below, profit, encoded by color intensity, has been added to the map.

FIGURE 6.64 Points that vary in size and color intensity can be used to display two quantitative variables at once.

In addition to using points that vary by size and color intensity, we can assign color intensities to the geographical regions themselves to encode aggregate values associated with those regions. Geographical displays that encode values in this way are called *choropleth maps*. In the following example, color intensities encode the populations of states.

FIGURE 6.65 Variations in color intensity can be used to assign values to geographical regions.

By combining all three means of encoding values on a map, we can display three quantitative variables simultaneously, as in the following example where the circle sizes represent sales revenues, the circle color intensities represent profit, and the state colors represent population.

FIGURE 6.66 Three quantitative variables can be associated with geographical regions at once.

In addition to associating quantitative values with specific locations or regions, it's sometimes useful to associate them with routes from place to place. For example, if you wanted to display varying amounts of traffic along specific roads, you might vary the thickness or the color intensity of the lines that represent those roads on the map. In the following example, the route and strength of each large storm in the Atlantic Ocean during the 2005 hurricane season is recorded using lines that vary in thickness.

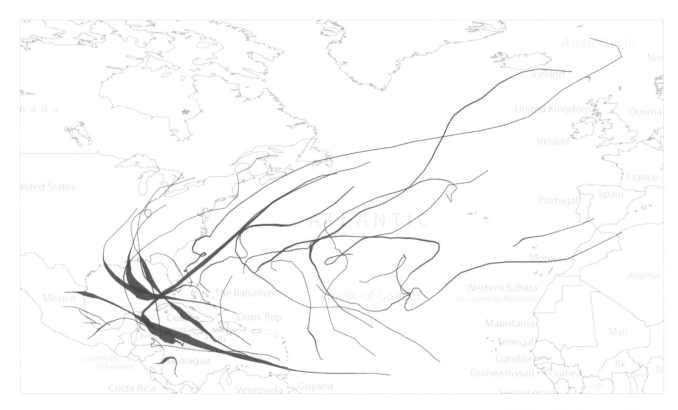

To summarize, we've learned that values can be displayed geospatially using the following means:

- Points of varying size (to pinpoint specific locations or for entire regions)
- Points or areas of varying color intensity (to pinpoint specific locations or for entire regions)
- Lines of varying thickness or color intensity (to display values associated with routes)

FIGURE 6.67 Varying line widths can associate values with geographical routes. This example was created by Richard Wesley and Chris Stolte of Tableau Software.

Summary at a Glance

Relationship	Value-Encoding Objects			
	Points	Lines	Bars	Boxes
Nominal Comparison	In the form of a dot plot when you can't use bars because the quantitative scale does not begin at zero	Avoid	Horizontal or vertical	Avoid
Time Series	In the form of a dot plot, but only when values were not collected at consistent intervals of time	Emphasis on overall pattern; categorical items on X axis, quantitative values on Y axis	Emphasis on individual values; categorical items on X axis, quantitative values on Y axis	Only when showing distributions as they change through time; categorical items on X axis, quantitative values on Y axis
Ranking	In the form of a dot plot, especially when you can't use bars because the quantitative scale does not begin at zero	Avoid	Horizontal or vertical	Only when ranking multiple distributions; horizontal or vertical
Part-to-Whole	Avoid	To display how parts of a whole have changed through time	Horizontal or vertical	Avoid
Deviation	As a dot plot when the quantitative scale does not begin at zero	Useful when combined with a time series	Horizontal or vertical, but always vertical when combined with time series	Avoid
Distribution				
Single	Known as a strip plot; emphasis on individual values	Known as a frequency polygon; emphasis on overall pattern	Known as a histogram; emphasis on individual intervals	Avoid
Multiple	Known as a strip plot; emphasis on individual values	Known as a frequency polygon; limit to a few lines	Avoid	Known as a box plot
Correlation	Known as a scatter plot	Avoid	Horizontal or vertical, in the form of a table lens	Avoid
Geospatial	Vary point sizes to encode values	To mark routes	Avoid	Avoid

PRACTICE IN SELECTING TABLES AND GRAPHS

Learning requires practice. Through practice you reinforce what you've learned by embedding it more securely in your memory and strengthen your ability to make connections between the concepts we've examined and their application to the real world.

You may be tempted to skip this section of practice exercises, but I encourage you to take a few minutes to work through them. Taking these few extra minutes now to strengthen and deepen what you've learned may save you countless hours and a great deal of frustration over the course of your lifetime. This section consists of six scenarios. Each requires that you make choices to determine the most effective design for communicating a quantitative message. Your choices for this set of exercises involve the following:

- Should the message be presented in the form of a table or a graph?
- If a table, which kind of relationship should it display?
  - Between a single set of quantitative values and a single set of categorical subdivisions
  - Between a single set of quantitative values and the intersection of multiple categories
  - Between a single set of quantitative values and the intersection of hierarchical categories
  - Among a single set of quantitative values associated with multiple categorical subdivisions
  - Among multiple sets of quantitative values associated with the same categorical subdivision
- If a graph, which kind of relationship should it display?
  - Nominal comparison
  - Time-series
  - Ranking
  - Part-to-whole
  - Deviation
  - Distribution
  - Correlation
  - Geospatial
- If a graph, which object or combination of objects for encoding the quantitative values would work best?
  - Points
  - Lines
  - Bars
  - Boxes

Space has been provided in the right margin for your answers to these questions for each of the six scenarios. Do your best to think through each scenario and respond without going back to review the contents of the chapters. Allow yourself to struggle a bit first before reviewing chapters for answers that you can't remember immediately. You want to reach the point where the information resides in your head and is thoroughly understood.

Scenario #1

You are a financial analyst who works for the new Chief Financial Officer (CFO). You've spent the last month providing reports to help her become familiar with the company's financial state. She has come to believe that expenses are excessive, so she has scheduled a series of meetings, one with each department head, to discuss the problem and explore possible remedies.

She would like you to provide a single report that includes, by department, the headcount and expenses to date for the current quarter compared to budgeted headcount and expenses. This will give her the basic information that she'll need for each of the meetings. It's up to you to provide this in a manner that will serve her purpose most effectively.

Responses to Scenario #1:
Table or graph?

If a table, which kind?

If a graph, what kind of relationship?

If a graph, which graphical objects for quantitative encoding?

Anything else?

Scenario #2

You work as a product marketing manager for a company that manufactures and sells five distinct lines of software: 1) business productivity, 2) educational, 3) games, 4) programming, and 5) utilities. During the past five years, the relative amount that each has contributed to overall revenue has shifted. Five years ago your programming products were on top, but today they are dead last, and games are on top. As you examine the relative sales of each product line for each of the last five years, you notice a clear decline in the success of products that are more technical in nature (i.e., programming and utilities) and an increase in those that are geared toward entertainment.

You are preparing strategic recommendations for the company's five-year plan. To set the stage for your recommendation that the programming and utilities product lines be sold off and the games line be expanded, you need to clearly present the shift that you've observed during the past five years. What form will you use to present this observation?

Responses to Scenario #2:
Table or graph?

If a table, which kind?

If a graph, what kind of relationship?

If a graph, which graphical objects for quantitative encoding?

Anything else?

Scenario #3

Six months ago you developed and began teaching a new course entitled "Ethical Management." When you initially proposed the idea for the class, your director was a little apprehensive about how well it would be received, but your past successes encouraged him to give you a shot. Now that you've been teaching it for a while and have worked the bugs out, it's time to give your director some evidence that he made the right decision in trusting your judgment and ability.

You've taught the course four times during the past month to a total of 100 students. Each student filled out an evaluation form at the end of the class, and you've tabulated the results. On a rating scale of 1 to 5, with 1 representing worthless and 5 representing excellent, the median rating for the course is 4, and the mean is 4.3. These ratings are exceptional. Not only is the average rating high, the range of ratings is tightly grouped around the ratings of 3, 4, and 5, with very few ratings of 2 and none of 1. When you compare these ratings to those that you received for another popular class that you also teach, their averages were about the same, but the spread of ratings for this other class were more broadly distributed, indicating that it doesn't work for all students as well as your new course.

You want to give this information to your director in a form that he will grasp with little difficulty. Once before, when you tried to communicate differences in the range of ratings between classes using standard deviations, you could tell that the director didn't really understand how to interpret them but was too embarrassed to admit it. This time you're going to approach the task differently. What form will your presentation take?

Responses to Scenario #3:
Table or graph?

If a table, which kind?

If a graph, what kind of relationship?

If a graph, which graphical objects for quantitative encoding?

Anything else?

Scenario #4

You have been promoted from Director of Customer Service to Vice President of Services. Before you were able to move full time into your new position, you had to recruit someone to replace you as director.

Your company spreads the work of customer service across four different customer service centers, one in each of four major geographical regions. Customers are able to rate their experiences with the service centers by responding to surveys distributed via email. Because you want the new director to focus on improving the centers that are scoring lowest in customers' ratings, you need to provide her with the mean rating of service for each service center during the most recent quarter. In what form will you present these summarized ratings?

Responses to Scenario #4:
Table or graph?

If a table, which kind?

If a graph, what kind of relationship?

If a graph, which graphical objects for quantitative encoding?

Anything else?

Scenario #5

You've been given a contract by a large manufacturing facility to analyze worker productivity data to see whether you can identify the cause of a recent decrease in productivity. What you learn from the new Operations Manager is that no matter how many additional people he hires, the result is reduced productivity. When the Operations Manager was hired six months ago, the General Manager told him that productivity had remained flat for years, and it was his job to increase it by 20% during the coming year. So far it has actually decreased by 20%.

After hearing this summary from the Operations Manager, one of the first things you decide to examine is the possible connection between staff additions and productivity decreases. Given your years of experience as a productivity analyst, you are not surprised to discover that increases in staff are proportionally related to decreases in productivity. You suspect that the addition of workers

Responses to Scenario #5:
Table or graph?

If a table, which kind?

If a graph, what kind of relationship?

If a graph, which graphical objects for quantitative encoding?

Anything else?

without changing anything else about the manufacturing process or facilities may have resulted in people simply getting in each other's way.

You decide to show the Operations Manager the strength of this relationship of increased staff to decreased productivity before taking any further steps. You have daily headcount and productivity statistics for the last year. Both headcount and productivity remained fairly steady until just after the Operations Manager's arrival. In what form will you present your information?

Scenario #6

For the first time ever, your organization has built a database that contains comprehensive and reliable information about donations. Since it became available, you've been slicing and dicing the information in various ways, looking for answers to important questions that you've never before been able to investigate. One of your queries involved a list of every single donation for the past year, sorted by size in U.S. dollars from the biggest to the smallest. You took your list and divided it into 10 equal groups labeled "Over 90 – 100%" (i.e., the top 10%), "Over 80 – 90%", and so on, to the final one labeled "Over 0 – 10%". Next, you calculated the running percentage of total income associated with the donations, beginning with the largest and continuing all the way to the smallest. You were then able to easily see the amount of income that each group of donations contributed to overall income.

You were amazed to discover that the top 10% of your donations contributed 87% of your total revenue. After the top 10%, the income contribution of the remaining 90% of your donations dropped off dramatically, with the last 50% contributing only 1% of total income. You have no doubt that your organization's leadership will find this discovery enlightening. You want to present this message as concisely and clearly as possible. You realize that if you don't hit them between the eyes with this important revelation in a single page of information, they won't bother reading it. What form will you give to this information to ensure that it hits the mark?

Responses to Scenario #6:
Table or graph?

If a table, which kind?

If a graph, what kind of relationship?

If a graph, which graphical objects for quantitative encoding?

Anything else?

You can find answers to the six scenarios in Appendix F, *Answers to Practice in Selecting Tables and Graphs.*

7 GENERAL DESIGN FOR COMMUNICATION

With a basic understanding of visual perception, we can build a set of visual design principles, beginning with those that apply equally to tables and graphs. Our primary visual design objectives will be to present content to readers in a manner that highlights what's important, arranges it for clarity, and leads them through it in the sequence that tells the story best.

Visual design can serve many purposes, not least of which is to create beauty, which we can appreciate purely for its own sake. This is the work of the artist. Without it our lives would be dismal and our souls malnourished. Artists spend their lives learning from the masters and their own painstaking experience. Through each stroke of the brush, angle of the chisel, or subtle positioning of the light, they attempt to move us in some way. As creators of tables and graphs, our use of visual design serves a different purpose but one that is also fundamental to life and deserves no less attention. We use visual design to communicate. There are stories in the numbers that will be perceived and acted upon or will go unnoticed and be ignored, depending on our knowledge of visual design and our ability to apply that knowledge to the important task of communication.

In this chapter we'll examine the aspects of visual design that apply equally to all visual forms for communicating quantitative information, including tables, graphs, and text. These general practices of communication-oriented design support two fundamental objectives:

1. Highlight
2. Organize

We highlight important information to give it a voice that comes through loudly and clearly, without distraction. We organize information to lead readers through it in a manner that promotes optimal understanding and use.

Highlight

It is appropriate to begin this section by repeating six incisive words written by Edward Tufte: "Above all else show the data."[1] These words should be our mantra. Nothing is more central to our task.

Tufte introduced a useful concept known as the data-ink ratio. Tables and graphs are composed of ink on the page. Some ink represents actual information and some does not (e.g., supporting components like grid lines or superfluous components like ornamentation that play no role whatsoever in presenting the data). The data-ink ratio is the amount of ink that presents information compared to the total amount of ink. The degree to which we feature data in a table

1. Edward R. Tufte (2001) *The Visual Display of Quantitative Information,* Second Edition. Graphics Press, page 92.

or graph corresponds in large part to the percentage of ink that we used to represent data rather than non-data.

The object isn't to eliminate all non-data ink. To some degree we always need supporting visual components to make tables and graphs readable. The object is to reduce the non-data ink to no more than what's necessary to make the data ink understandable.

We highlight data through a design process that involves activities of two types:

1. Reducing the non-data ink
2. Enhancing the data ink

Reduce the Non-Data Ink

The process of reducing the non-data ink involves two steps:

1. Subtract unnecessary non-data ink
2. De-emphasize and regularize the remaining non-data ink

SUBTRACT UNNECESSARY NON-DATA INK

Subtracting unnecessary non-data ink begins by asking the following question about each visual component: "Would the data suffer any loss of meaning or impact if this were eliminated?" If the answer is "no," then get rid of it. Resist the temptation to keep things just because they're cute or because you worked so hard to create them. If they don't support the message, they don't serve the purpose of communication. As the author Antoine de Saint-Exupery suggests: "In anything at all, perfection is finally attained not when there is no longer anything to add, but when there is no longer anything to take away."[2]

By subtracting what is not needed to support the message, you bring your communication one step closer to elegance. The word elegance comes originally from the Latin term *eligere*, which means to choose out or to select carefully. To achieve elegance in communication, you must carefully select the content that is essential to the message and trim all else away.

DE-EMPHASIZE AND REGULARIZE THE REMAINING NON-DATA INK

Once you've subtracted all the unnecessary non-data ink, you should push the non-data ink that remains far enough into the background to enable the data to stand out clearly in the foreground. This can be achieved by reducing the visual prominence of non-data ink components.

Tables and graphs consist of three visual layers: 1) *data* as the top or prominent layer, 2) *non-data items* as the middle layer, and 3) the *background* (the surface on which the data and supporting components reside). Non-data items, consistent with their supporting role, should stand out just enough from the background to serve their purpose but not so much that they draw attention to themselves. This can be achieved through the use of thin lines and soft, neutral colors (e.g., light gray). To do otherwise, for example to give the same visual weight to data and non-data items, would provide no visual cues to lead the reader's eyes to what's important. When everything stands out, nothing stands out.

2. This quotation of Antoine de Saint-Exupery and the explanation of the term elegance were taken from Kevin Mullet and Darrel Sano (1995) *Designing Visual Interfaces*. Sun Microsystems, Inc., page 17.

Because a reader's eyes are drawn to contrast, you can go one step further to reduce the visibility of non-data ink by making it as consistent as possible, so that none of it stands out. Multiple instances of the same supporting components throughout a report should look precisely the same everywhere they appear. Any differences work against your purpose by inviting your readers' eyes to notice and their brains to assign meaning to those differences.

Take a few minutes now to examine two or three of your own reports to identify opportunities to reduce the non-data ink. You may be surprised to find how much there is that could be subtracted, muted, and regularized for greater effect.

. . . .

Enhance the Data Ink

You can enhance the data ink through a process that consists of two steps:

1. Subtract unnecessary data ink
2. Emphasize the most important data ink

SUBTRACT UNNECESSARY DATA INK

You must carefully avoid the common mistake of saying too much. Not all information is equally important. This is especially true when your readers don't have the time or the patience necessary to savor a message in all its subtlety. Don't remove anything that's important, but be sure to remove all that is peripheral to the interests and purposes of your readers. Every step taken to reduce data causes what remains to stand out even more. The more you earn your readers' trust by giving them only what they need, the more they'll pay attention to everything you give them.

The intention here is to summarize when detail isn't necessary and to trim away what's not important, not to arbitrarily reduce the content of your message. It's appropriate for a single table or graph to deliver a great deal of information or to articulate a complex (but not overly complicated) message. Strive to give your readers what they need, and all that they need, but nothing more.

EMPHASIZE THE MOST IMPORTANT DATA INK

Data values are encoded differently in tables than in graphs. In tables, they are encoded entirely in verbal language (i.e., words and numbers), but in graphs they are encoded primarily in visual language (e.g., points, lines, bars, and boxes) although words and numbers are used as well. Regardless of the encoding method, certain visual attributes of objects, words, and numbers stand out more than others.

In the earlier chapter on visual perception, you learned that preattentive visual attributes differ in the degree to which they stand out. Size is a good example. You can make something stand out by making it bigger. Objects, words, and numbers that are bigger stand out more than those that are smaller, all else being equal. You can take advantage of this to emphasize the most important data ink relative to the rest. Here is a list of the preattentive visual attributes that are especially useful for emphasizing data ink in tables and graphs:

Attribute	Values Useful for Emphasis
Width	Thicker lines (including words and numbers that are boldfaced) stand out more than thinner lines.
Orientation	Slanted words and numbers (i.e., italics) stand out more than those that are oriented normally (i.e., not slanted), assuming that vertically oriented fonts are the norm.
Size	Bigger objects, words, and numbers stand out more than smaller objects.
Enclosure	Objects, words, and numbers that are enclosed by lines or background fill color stand out more than those that are not enclosed.
Hue	Objects, words, and numbers that have a hue that is different from the norm stand out.
Color intensity	Objects, words, and numbers that are dark or bright stand out more than those that are light or pale.

Each step in the process of highlighting data results in greater simplicity. In the communication of quantitative information, simplicity of design is the essence of elegance. Your message might be complex, but its design—the form in which you present it—should be so simple that to your readers it is nearly invisible.

Organize

When your readers look at a page or screen of information, they immediately begin to organize what they see in an effort to make sense of it. As a designer of communication, it is your job to organize the information for them in a manner that tells the story as clearly as possible. If you fail to do this effectively, the result is not information that is unorganized, but information that is organized in a manner that does not support its essential story, resulting in ineffective communication. In fact, your readers may get a different message entirely—perhaps one that is wrong. Communication involves much more than knowing what to say; it also involves knowing how to say it.

The page or screen that serves as your medium of communication will often contain more than a single table or graph. Your message may require multiple tables, multiple graphs, or a combination of both, along with additional text in the form of annotations, sentences, or even whole paragraphs. When you arrange information on the page or screen, you must consciously do so to tell a story. What should I say first? What should I save for last? What should I emphasize more than the rest? The answers to these questions take on the form of visual attributes designed to accomplish the following:

1. Group (i.e., segment information into meaningful sections)
2. Prioritize (i.e., rank information by importance)
3. Sequence (i.e., provide direction for the order in which information should be read)

Group

You must always begin with a clear sense of what belongs together, that is, what your readers should perceive as belonging to the same group because those units of information have something in common. Once this is clear, you can select from various visual design techniques that organize the information into groups.

Grouping takes place on several levels. You begin with your overall message and then break down its content into different topics. The various topics are then grouped into the appropriate modes of expression: tables, graphs, and text. Within tables and graphs, information is naturally grouped into categorical and quantitative data. Finally, categorical information is broken down into its various items, and quantitative values are associated with each of those categorical items.

It's your job to make this grouping obvious to your readers. It shouldn't be up to your readers to do the work of arranging the content into meaningful groups when you can do this in advance for them.

The Gestalt principles of visual perception reveal a number of techniques that can be used to group information meaningfully. The simplest approach—proximity—is often the best. This is especially true for arranging content into various topics. If you were communicating quarter-to-date sales performance to a sales manager, your overall message might consist of regional sales performance compared to forecast, the top orders, and the top customers. You would have a single story consisting of three related topics. By placing the information related to each topic close together and by separating the topic groups by white space, you create a simple and clear arrangement with nothing to hinder your readers' eyes as they move from one group to another.

Sometimes your message consists of several separate but related topics that need to be appropriately grouped and arranged on a page or screen, and among those individual topics reside relationships that should be identified. Let's continue the previous example. The story of quarter-to-date sales consists of three primary topics, but each includes sales expressed both as bookings and billings. Let's assume that you have appropriately arranged the three topics into a graph that displays regional sales performance compared to the forecast, a table listing the top orders, and a table listing the top customers, each separated by enough white space to render it distinct. Even though bookings and billings both appear in the graph as well as the tables, it would be helpful to tie the separate instances of each together visually. This would make it easier for your readers to quickly scan for all bookings information separately from billings information, and vice versa. You could do so simply by selecting one of the remaining Gestalt principles or one of the preattentive attributes that you learned about in the chapter on visual perception. Given this scenario, what attribute or principle might you choose to visually group bookings as distinct from billings? Take a moment to run through the list of available methods, weighing the potential advantages of each.

．．．．．．．

There is no one right answer. You might have realized while assessing the alternatives, however, that many of the available methods suggest that either bookings or billings are more important than the other. For instance, if you chose color intensity to distinguish bookings from billings, rendering bookings as black and billings as gray, bookings would stand out more. This would be appropriate if the purpose of your message were to emphasize bookings over billings, but if you wished to treat them equally, variation in color intensity would not be the best method. You could, however, select different hues for each.

Tables and graphs both use conventional means to organize information into categories. Tables primarily use the Gestalt principles of proximity and continuity to organize different categories into columns and rows. Graphs use many techniques, such as the principles of similarity (e.g., common hues or shapes) and connection (e.g., the use of a line to connect points) to group data. We'll examine these techniques further in the coming chapters on table and graph design.

Prioritize

Whenever you communicate quantitative information, it is important to step back and ask yourself, "What are the important numbers here?" Once you've established a clear sense of what's important, you should make that information stand out clearly from the rest. This is a vital part of your job. Don't just highlight important numbers when you happen to think about it or when their importance hits you over the head. Consider it every time.

Not all numbers are equally important. In fact, some numbers are so much more important than others that a few seconds spent examining and understanding them produces benefits that could never be equaled by years of concentration on all the others. Help your readers develop a productive approach to numbers by pointing out those that most deserve attention.

We learned in the chapter on visual perception that some preattentive visual attributes are perceived quantitatively. Their values can be arranged along a continuum ranging from less to more, small to big. Such attributes have the built-in ability to make some information stand out as more prominent than the rest. Here's a reminder of the attributes that are perceived quantitatively, along with examples of how each can be used to highlight important sections of text, tables, and graphs:

Attribute	Tables and Text	Graphs and Objects
Width	• Boldfaced text	• Thicker graph lines • Wider bars
Size	• Bigger tables • Larger fonts	• Bigger graphs • Bigger symbol shapes
Color intensity	• Darker or brighter colors	• Darker or brighter colors
2-D position	• Positioned at the top • Positioned at the left • Positioned in the center	• Positioned at the top • Positioned at the left • Positioned in the center

As the final item in the list suggests, certain positions in a 2-D space stand out as more prominent than the rest. When language is read from left to right and top to bottom, you can make text, tables, or graphs appear more prominent by locating them at the top left of a page or screen.

In addition to these quantitatively perceived visual attributes, you can also take advantage of the fact that visual contrast of any kind can make particular information stand out from the rest. Here are additional attributes that you can use to emphasize particular information by means of contrast:

Attribute	Tables and Text	Graphs and Objects
Orientation	• Italics	• Data points with an orientation that is different from the norm
Shape	• Any font that is different from the norm	• Any symbol shape that is different from the norm
Enclosure	• Border around tables, rows, columns, or particular values • Fill color behind tables, rows, columns, or particular values	• Border around graphs or particular values • Fill color in a graphs behind particular values
Hue	• Almost any hue that is different from the norm	• Almost any hue that is different from the norm
2-D position	• Any position that is out of vertical or horizontal alignment with the norm	• Any position that is out of vertical or horizontal alignment with the norm

The final attribute of 2-D position highlights the significance of alignment in visual design. We are more sensitive to the vertical and horizontal alignment of text and objects than you might imagine. The slightest misalignment jumps out at us, and we react by trying to impose meaning on that difference. Unless you intend to make something stand out, be careful to keep the edges of text and objects aligned so that your readers' eyes can scan down and across without disruption.

You may use differences in horizontal alignment quite consciously to establish a hierarchical relationship between different sections of content, with subordinate content indented to the right of higher-level content. When you use indentation in this manner, be sure to indent far enough to make your intention clear. Using alignment in this manner makes it easy for your readers to separately scan higher-level content without distraction from subordinate content when they wish to take in the main points quickly.

There is actually one more method that we haven't considered yet because it doesn't involve a visual attribute of objects but is instead a special type of object that's used for a particular purpose. I'm referring to a collection of objects called *pointers*. This includes objects like arrows, asterisks, and check marks. Put one of these next to or pointing to any content, and your reader's attention will definitely be drawn to it. Pointers are not subtle, especially arrows, so you should use them with discretion to avoid visual clutter.

Sequence

The final objective of visual design is to provide clear direction to your readers regarding the best sequence in which to read a report's contents. The strongest sequencing method is the location of content in 2-D space. Because we read from left to right and top to bottom, this is generally the order in which your readers will scan the page or screen. If you clearly divide the contents into columns, such as those in newspapers, readers will first scan the left-most column from top-to-bottom, then move to the top of the next column, unless you've done something to draw their attention elsewhere.

The strength of this left-to-right and top-to-bottom navigational sequence is greatest with textual content because text can only be perceived through the sequential process of reading. This same sequence works for graphs as well but not as strongly. For instance, if your page contains a collection of graphs without sections of text to introduce them, readers will still give attention to each graph in the normal left-to-right, top-to-bottom sequence, all else being equal. However, if any one of the graphs has been highlighted as important using any of the prioritizing methods noted in the previous section, readers' eyes will likely be drawn immediately to that graph. If your message requires that your readers work their way through a collection of tables and/or graphs in a particular order, you can further reinforce the navigational route by using numbers (1, 2, 3, etc.), alphabetical letters (A, B, C, etc.), or some other form of sequential labeling.

Integrate Tables, Graphs, and Text

Tables, graphs, and text form a powerful team, working together intimately to communicate quantitative information. Each brings a different set of strengths to the effort. We've already examined the separate strengths of tables and graphs. In this section, we'll focus on the contribution of text and the way it can be integrated with tables and graphs to create clear and powerful messages.

The Role of Text

To complement or enhance tables and graphs, text can:

- Label
- Introduce
- Explain
- Reinforce
- Highlight
- Sequence
- Recommend
- Inquire

Let's take a quick look at each role.

LABEL

We've already examined the role of labeling. Tables and graphs both use text to label information. Tables use text (i.e., words and numbers) not only to express quantitative and categorical data but also to label what the columns and rows contain. Graphs incorporate text in the form of titles, labels for categorical items and quantitative values along the scale lines, and legends to interpret the visual encoding of categorical items (e.g., the orange bars represent the eastern sales region). Text in the form of labels supplies critical information to enable readers to interpret tables and graphs.

Reports containing tables and graphs also use text in the form of titles. Clear titles are vital data in themselves. How many times have you seen a report with a title that revealed nothing definitive about its contents? When people scan lists of available reports in an effort to find one that contains the information they need, they often do this with no information other than the titles. Good titles are invaluable.

INTRODUCE

Quantitative displays often require an introduction to set the reader on a clear path to understanding. Text is the ideal medium for providing introductions. Introductions are especially useful in new reports and for new readers of old reports, potentially saving readers a great deal of time and frustration. Among other things, an introduction should preview what readers will find in the report, what they should especially notice, and what they should do with the information. Because you can't always hand a report directly to all its eventual readers, the introduction is your chance to set the stage for the report using text that states what you would tell them in person if you could.

EXPLAIN

An introduction to a report is not the best place to put every bit of text that might be needed to explain the data the report contains. Explanations work most effectively when they are provided right where they're needed to clarify something about the message. If you provide a time-series graph that displays an unusual brief up-tick in donations during the month of May, you may want to mention right there, in or just underneath the graph, that a successful promotional campaign beginning in late April was responsible for the anomaly. If a few words are what it takes to make the message clear, then they belong there. Whenever a table or graph doesn't speak clearly enough on its own, its design should be improved to solve the problem, if possible, or a little text should be added.

REINFORCE

Some information is so important that you should say it more than once and in more than one way to increase its likelihood of getting through to your readers. If you encode that information visually in a graph or verbally in the columns and rows of a table, and then present it again in a few well-chosen words, you will increase the odds that the message will be heard. You don't want to overdo it though. Don't say everything in multiple ways or you'll waste your readers' time and lose their confidence. The important stuff, however, deserves a little extra.

HIGHLIGHT

We've examined several methods for visually highlighting important data. Sometimes it's also useful to highlight particular information by referencing it with words as well. This is different from reinforcement because in this case you're not repeating the information in a different form; you're simply calling the reader's attention to it. For instance, if the sales ranking of a particular product warrants special notice, you may say so in words right in or underneath the graph. Perhaps it isn't appropriate to make that product stand out above the rest visually in the graph itself because that would distract from the other products that are also important, but a short note following the graph could do the job without creating a visual distraction in the graph.

SEQUENCE

Sometimes it's challenging to use visual methods alone to clearly reveal the order in which your readers should examine the contents of a report. Perhaps information in a report cannot be positioned from left to right and top to bottom in the order it should be read because of a greater need to use 2-D location to highlight the importance of some data or to group data in a particular way. In circumstances such as these, you can use text to instruct your readers to navigate through the contents in a particular way.

RECOMMEND

As a communicator of important quantitative information, your job often involves more than simply informing. Sometimes it's your responsibility to recommend what could or should be done. Recommendations for action are best communicated in words. Whether or not making recommendations is your explicit role, your organization might appreciate it if you take the initiative to offer recommendations that you deem warranted.

INQUIRE

Inquiry is vastly underrated and too often ignored. Quantitative information frequently invites questions that ought to be asked. You can sometimes add more value to your organization by asking a single important question than by providing hundreds of answers. We so often get caught up in business as usual that we fail to question why things are as they are or whether things might be better if they were different. As a communicator of quantitative information, you're in a great position to recognize opportunities for further exploration, important speculation, and valuable questions that somebody ought to be considering. Why not ask such questions by placing a few words in your reports near the information that prompted the questions? I realize that your readers might respond by assigning you the task of exploring those questions, but who better? If you're like me, discovering the right questions to ask and then doing the research and analysis to find the right answers is the real fun of working with numbers.

Text Placement

Tables, graphs, and text are complementary. There is no need to arbitrarily place them in separate areas in a report. Blend them together, placing each unit of content precisely where it is most needed. Just be careful, when you place text in the plot area of a graph, to do so in a way that does not obscure or distract from patterns in the data.

The importance of minimizing distraction was suggested indirectly in the chapter about visual perception. Our eyes have a limited area on which they can focus at any one time. If you place the legend for a graph too far from the data it labels, you force your readers to jump back and forth over and over to read the graph because they can't keep all the encodings (e.g., the blue line represents widgets) in working memory. If you place the explanation for a table that appears on page 1 at the end of the report on page 10, you'll cause unnecessary effort and frustration. Perhaps your message involves a great deal of text spread across several pages, which refers to a single table or graph. In this case, you might actually want to reproduce the table or graph in multiple locations so that it's always available where it's needed.

You might have noticed that in this book I don't follow the traditional practice of placing notes and references at the bottom of the page, the end of the chapter, or worst of all, the end of the book. I also don't force you to turn to a middle section of illustrations but instead have integrated all illustrations right where they're needed. This was a conscious design choice to support your reading experience. You face similar design choices regarding the integration of tables, graphs, and text whenever you construct a quantitative message. The tighter the integration, the better.

Required Text

Text should be included on every page of every report to answer the following questions:

- What?
- When?
- Who?
- Where?

Excerpts from multi-page reports are often copied and distributed. If the information that identifies the report only appears at the beginning, readers who have only a portion of the report will have no way of knowing where it came from. It takes only a minute to include this identifying information in the page header or footer of your reports.

WHAT

As I said before, a good title is invaluable. A simple glance at the title should clearly tell your readers what the report contains. The title should describe, without being long winded, the type of quantitative and categorical information that the report presents. The title "Sales" isn't enough. How is this report different from all the other reports that deal with sales? A title like "2011 Bookings by Month and Region" says a great deal more.

WHEN

Two facts should be provided with every report to inform your readers about its relation to time:

- The range of dates the information represents
- The point in time when the information was collected

Does the information represent a single hour, day, week, month, quarter, year? Does it represent some range of hours, days, etc.? Perhaps it represents an odd range of time, such as from April 23rd of 2010 through January 14th of 2011. Whatever the range, if it isn't clearly labeled in the table or graph, then make sure it appears in the title or subtitle.

The point in time when the information was collected is often called the "as of" date. "This represents expenses for February as of March 4th." This information is important because more expenses could be recorded later, or corrections could still be made to expenses after March 4th. Multiple reports covering the same period of time often differ simply because they were produced at different times, and the data changed in between them. A simple "as of" followed by the date when the information was collected, noted in the header or footer of the report, conveniently satisfies this need.

WHO

The reason to include your name or the name of the group you represent on your reports is not self-promotion; it is to let people know whom to contact if they have questions. I've spent many frustrating hours during the course of my career trying to track down the creator of a report because I needed to ask a simple question about it. Save your readers this annoyance. Provide your name, along with some means to contact you, such as an email address or phone number.

WHERE

By "where" I am referring to page numbers, which tell your readers where they are in a multi-page report. Try describing to someone where he or she can find a particular piece of information in a multi-page report that doesn't include page numbers. I find that the format "Page # of ##" (e.g., "Page 13 of 197") is best because it informs your readers from the very first page how many pages they're facing in total. This is especially helpful when reports are distributed and read electronically because there is no physical stack of pages to alert readers to the size of the report. Have you ever started to print an electronic report only to realize later when you saw the line of angry coworkers at the printer that it was more than a thousand pages long?

Summary at a Glance

General Design Objectives of Quantitative Communication

HIGHLIGHT

- Reduce the non-data ink
  - Subtract unnecessary non-data ink
  - De-emphasize and regularize the remaining non-data ink
- Enhance the data ink
  - Subtract unnecessary data ink
  - Emphasize the remaining data ink

ORGANIZE

- Group
- Prioritize
- Sequence

Highlight What's Important

USING QUANTITATIVELY PERCEIVED VISUAL ATTRIBUTES

Attribute	Tables and Text	Graphs and Objects
Width	• Boldfaced text	• Thicker graph lines • Wider bars
Size	• Bigger tables • Larger fonts	• Bigger graphs • Bigger symbol shapes
Color intensity	• Brighter, more vivid colors	• Brighter, more vivid colors
2-D position	• Positioned at the top, left, or center	• Positioned at the top, left, or center

USING VISUAL ATTRIBUTES IN CONTRAST TO THE NORM

Attribute	Tables and Text	Graphs and Objects
Orientation	• Italics	• Data points with an orientation that is different from the norm
Shape	• Any font that is different from the norm	• Any symbol shape that is different from the norm
Enclosure	• Border around or shading behind table, rows, columns, or particular values	• Border around or shading behind graph or particular values
Hue	• Almost any hue that is different from the norm	• Almost any hue that is different from the norm
2-D position	• Any position that is out of vertical or horizontal alignment with the norm	• Any position that is out of vertical or horizontal alignment with the norm

Sequence Information
- Using left-to-right, top-to-bottom positioning
- Using visual highlighting
- Using sequential labels (e.g., 1, 2, 3 . . .)

Include on Every Page

What it is	In the form of a good title
When it is	In the form of the range of dates and an "as of" date
Who produced it	So readers know whom to contact
Where readers are	In the form of page numbers

8   TABLE DESIGN

Once you've determined that a table should be used to communicate your message and have chosen the type of table that will work best, you should refine your design so that it can be quickly and accurately read and understood.

We use tables regularly to communicate lists of quantitative information. They've been around since the 2nd century common era (CE) and have become commonplace since the advent of spreadsheet software in the 1970s. When used properly and designed well, they're fantastic, but they often fall far short of their potential. Fortunately, the design practices required to optimize their effectiveness are simple and easy to learn.

Structural Components of Tables

The components that we combine to construct tables and graphs fall into two categories:

- Information components
- Support components

Each component, whether its role is to express information directly or to play a supporting role by arranging or highlighting information, must be thoughtfully designed to do its job effectively. When these components are designed well, the result is clear, accurate, and efficient communication.

Data Components

Tables encode information as text (i.e., words and numbers). They include both categorical items and quantitative values. Tables generally work best when they also contain additional text that is used to complement categorical items and quantitative values in various ways. In Chapter 7, *General Design for Graphical Communication*, we noted that additional text can do the following:

- Label
- Introduce
- Explain
- Reinforce
- Highlight
- Sequence
- Recommend
- Inquire

Apart from these complementary roles, the most common uses of text in tables are as titles and headers. Headers are used to label columns of data (column headers) and rows of data (row or group headers).

In total, then, tables actually consist of three types of information:

- Categorical labels
- Quantitative values
- Complementary text

Support Components

Support components are the non-data ink objects that highlight or organize the data. The following visual objects and attributes can function as support components:

- White space and page breaks
- Rules and grids
- Fill color

WHITE SPACE AND PAGE BREAKS

The defining structural characteristic of tables is the arrangement of information into columns and rows. The primary visual means that we can use to impose this structure is *white space.*

White space is used to group data objects that belong together by separating them from others. It is the visual mechanism that underlies the Gestalt principle of proximity. White space is not the most visually powerful means of grouping objects (e.g., the Gestalt principle of enclosure is more powerful), but its subtlety and lack of ink make it especially useful. Properly used, white space clearly organizes data objects into groups without drawing attention to itself. Any design method that works effectively without drawing attention away from the information is invaluable.

Page breaks are a logical extension of white space. They can be used to group data by starting each new group on a separate page or screen. Once again, the underlying principle is proximity. Page breaks should not be used to unnecessarily separate information that ought to be seen together, however. For example, a page break would not be useful to separate groups of information that are meant to be compared.

Technically, the term "white space" is a bit of a misnomer in that the space it defines is only white if the color of the background is white. The alternate term "blank space" is technically more accurate, but we'll stick with the conventional term to keep things simple.

RULES AND GRIDS

Rules and grids are variations on the same theme. Both use lines to delineate or highlight data. Grids are combinations of horizontal and vertical lines that intersect to form a matrix of rectangles around data. Rules are lines that run either only horizontally or only vertically and therefore don't intersect. Grids work on the Gestalt principle of enclosure, and rules work primarily on the principle of connection even though they do not actually touch the data that they tie together.

FILL COLOR

2-D areas containing information in a table can be filled with different color intensities, such as shades of gray, or different hues to group areas of the table together or highlight them as important. Both use the Gestalt principles of similarity and enclosure to define an area within a table (e.g., one or more columns and/or rows) and make the area stand out.

The surface on which all data and support components reside is the background. It provides a clear surface that allows information, as well as the support components that arrange and highlight the information, to stand out for easy and efficient reading.

Visual Attributes of Components

All components of tables (data and support) exhibit visual attributes. Design involves not only choosing the components that are used to construct the table and display the information, but also determining the visual attributes of each. Throughout the rest of this chapter, we'll explore the best choices for specific situations.

Table Terminology

It is helpful, when discussing table design, to begin with a standard set of terms for the parts of a table. *Figure 8.1* shows the terms that we'll use throughout this chapter. Let's clarify terms that might be new or ambiguous. In the context of tables that contain quantitative information, the term *body* generally refers exclusively to the rectangular area that contains the quantitative values. Only the area that is shaded in light gray constitutes the body of the sample table in *Figure 8.1*. Defining body in this manner highlights the centrality of the quantitative information, reinforcing the fact that the primary message is in the numbers.

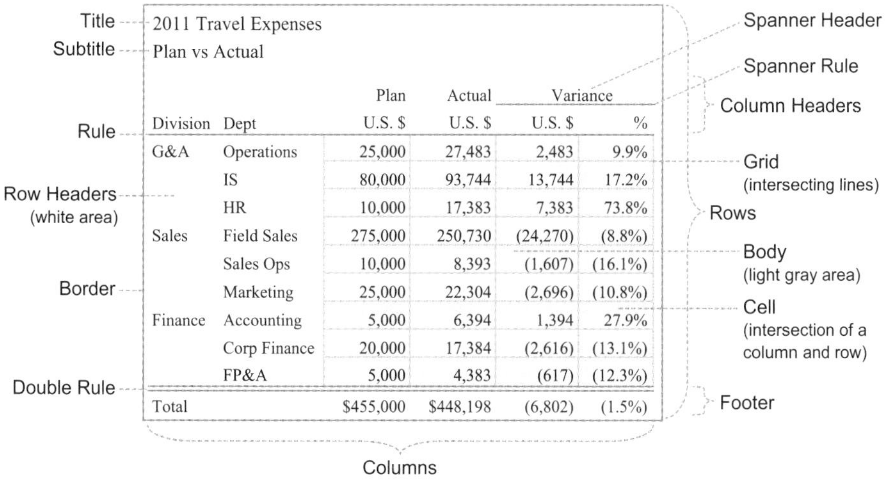

2011 Travel Expenses
Plan vs Actual

		Plan	Actual	Variance	
Division	Dept	U.S. $	U.S. $	U.S. $	%
G&A	Operations	25,000	27,483	2,483	9.9%
	IS	80,000	93,744	13,744	17.2%
	HR	10,000	17,383	7,383	73.8%
Sales	Field Sales	275,000	250,730	(24,270)	(8.8%)
	Sales Ops	10,000	8,393	(1,607)	(16.1%)
	Marketing	25,000	22,304	(2,696)	(10.8%)
Finance	Accounting	5,000	6,394	1,394	27.9%
	Corp Finance	20,000	17,384	(2,616)	(13.1%)
	FP&A	5,000	4,383	(617)	(12.3%)
Total		$455,000	$448,198	(6,802)	(1.5%)

FIGURE 8.1 This diagram labels the parts of a table.

This figure is meant to label a table's components, not to illustrate best practices.

Rows that summarize information from preceding rows are called *footers*. These can be used to summarize the entire table, as in the diagram above, or to summarize a subset of rows, in which case they are called *group footers*. The two left-most columns in the table above contain categorical items. When categorical items appear as labels to the left of the quantitative values associated with them—a common arrangement—they serve as labels for the rows and are therefore called *row headers*. When labels appear above the column of data associated with them, they are *column headers*. When a column header spans multiple columns, it is called a *spanner header*. When a rule is used to underscore a spanner header and does so by spanning each of the columns to which the header applies, it is called a *spanner rule*.

Table Design Best Practices

In this section we'll dig into the details of table design. We'll cover several topics, grouped into five categories:

- Delineating columns and rows
  - White space
  - Rules and grids
  - Fill color
- Arranging data
  - Columns or rows
  - Groups and breaks
  - Column sequence
  - Data sequence
- Formatting text
  - Orientation
  - Alignment
  - Number and date format
  - Number and date precision
  - Fonts
  - Emphasis and color
- Summarizing values
  - Column and row summaries
  - Group summary values
  - Headers versus footers
- Page information
  - Repeating column headers
  - Repeating row headers

Delineating Columns and Rows

The table design process involves several decisions regarding the layout of columns and rows to provide a structure that is easy and efficient to read and understand. Readers should be able to quickly scan through the content to find what they need and perhaps make localized comparisons of related numbers.

WHITE SPACE

White space is the preferred means for arranging data into columns and rows. The subtle use of blank space to group data into columns and rows is the least visible means available. It does its job without calling attention to itself.

The ability of white space to effectively delineate columns and rows is only limited when the overall space available for the table is so restricted that white space alone can't keep the columns or rows sufficiently distinct. If two rows of data are too close together, our eyes are not able to easily track across one without confusing it with the other. The same is true, but to a lesser extent, with columns.

As rows grow wider across, you need more white space between them to enable your readers' eyes to track across them without difficulty. When faced with wide rows, you have two potential means to address them using white space alone: 1) Decrease the horizontal white space between the columns to clarify the continuity of the rows, or 2) increase the vertical white space between the rows. Obviously, you can only decrease the horizontal white space between the columns so much before you reach the point where the columns are no longer clearly distinct. You can then add more white space between the rows, but too much can result in too little data on the page or screen. To preserve data density, you sometimes need to add a visual component other than white space (e.g., rules) to separate the rows. The balance between white space and overall data density is upset when the vertical white space between the rows exceeds the vertical space used by the rows of data themselves. Let me illustrate. When I created *Figure 8.2* using Excel, I selected a 10-point font (i.e., one that is 10 points high) and a row height of 12 points, resulting in 2 points of vertical white space between the rows (i.e., 20% of the height of the data). Given a full 12 months of data spread horizontally across the page, it is somewhat difficult to track across the rows with this spacing.

Product	Jan	Feb	Mar	Apr	May	Jun	Jul	Aug	Sep	Oct	Nov	Dec
Product 01	93,993	84,773	88,833	95,838	93,874	83,994	84,759	92,738	93,728	93,972	93,772	99,837
Product 02	87,413	78,839	82,615	89,129	87,303	78,114	78,826	86,246	87,167	87,394	87,208	92,848
Product 03	90,036	81,204	85,093	91,803	89,922	80,458	81,191	88,834	89,782	90,016	89,824	95,634
Product 04	92,737	83,640	87,646	94,557	92,620	82,872	83,626	91,499	92,476	92,716	92,519	98,503
Product 05	86,245	77,785	81,511	87,938	86,136	77,071	77,773	85,094	86,002	86,226	86,043	91,608
Product 06	88,833	80,119	83,956	90,576	88,720	79,383	80,106	87,647	88,582	88,813	88,624	94,356
Product 07	82,614	74,511	78,079	84,236	82,510	73,826	74,498	81,511	82,382	82,596	82,420	87,751
Product 08	85,093	76,746	80,421	86,763	84,985	76,041	76,733	83,957	84,853	85,074	84,893	90,384
Product 09	87,646	79,048	82,834	89,366	87,535	78,322	79,035	86,475	87,399	87,626	87,440	93,095
Product 10	90,275	81,420	85,319	92,047	90,161	80,672	81,406	89,070	90,021	90,255	90,063	95,888

FIGURE 8.2 This example shows rows of data without enough white space between them, which makes horizontal tracking difficult.

In the next example below, the same information is displayed with white space between the rows equal to the height of the rows.

Product	Jan	Feb	Mar	Apr	May	Jun	Jul	Aug	Sep	Oct	Nov	Dec
Product 01	93,993	84,773	88,833	95,838	93,874	83,994	84,759	92,738	93,728	93,972	93,772	99,837
Product 02	87,413	78,839	82,615	89,129	87,303	78,114	78,826	86,246	87,167	87,394	87,208	92,848
Product 03	90,036	81,204	85,093	91,803	89,922	80,458	81,191	88,834	89,782	90,016	89,824	95,634
Product 04	92,737	83,640	87,646	94,557	92,620	82,872	83,626	91,499	92,476	92,716	92,519	98,503
Product 05	86,245	77,785	81,511	87,938	86,136	77,071	77,773	85,094	86,002	86,226	86,043	91,608
Product 06	88,833	80,119	83,956	90,576	88,720	79,383	80,106	87,647	88,582	88,813	88,624	94,356
Product 07	82,614	74,511	78,079	84,236	82,510	73,826	74,498	81,511	82,382	82,596	82,420	87,751
Product 08	85,093	76,746	80,421	86,763	84,985	76,041	76,733	83,957	84,853	85,074	84,893	90,384
Product 09	87,646	79,048	82,834	89,366	87,535	78,322	79,035	86,475	87,399	87,626	87,440	93,095
Product 10	90,275	81,420	85,319	92,047	90,161	80,672	81,406	89,070	90,021	90,255	90,063	95,888

FIGURE 8.3 This example shows rows of data with enough white space between them so that horizontal tracking is made easy.

This is the practical limit: a one-to-one ratio of data height to white space height. Notice that your eyes track across these rows with little difficulty. In fact, even a little less white space between the rows would still work effectively.

In the next example, I've exceeded the practical limits of vertical white space, with white space that equals 150% of the height of the data. In this case you certainly don't have any difficulty tracking across the rows, but too much space is wasted, reducing data density to an unacceptable degree.

Product	Jan	Feb	Mar	Apr	May	Jun	Jul	Aug	Sep	Oct	Nov	Dec
Product 01	93,993	84,773	88,833	95,838	93,874	83,994	84,759	92,738	93,728	93,972	93,772	99,837
Product 02	87,413	78,839	82,615	89,129	87,303	78,114	78,826	86,246	87,167	87,394	87,208	92,848
Product 03	90,036	81,204	85,093	91,803	89,922	80,458	81,191	88,834	89,782	90,016	89,824	95,634
Product 04	92,737	83,640	87,646	94,557	92,620	82,872	83,626	91,499	92,476	92,716	92,519	98,503
Product 05	86,245	77,785	81,511	87,938	86,136	77,071	77,773	85,094	86,002	86,226	86,043	91,608
Product 06	88,833	80,119	83,956	90,576	88,720	79,383	80,106	87,647	88,582	88,813	88,624	94,356
Product 07	82,614	74,511	78,079	84,236	82,510	73,826	74,498	81,511	82,382	82,596	82,420	87,751
Product 08	85,093	76,746	80,421	86,763	84,985	76,041	76,733	83,957	84,853	85,074	84,893	90,384
Product 09	87,646	79,048	82,834	89,366	87,535	78,322	79,035	86,475	87,399	87,626	87,440	93,095
Product 10	90,275	81,420	85,319	92,047	90,161	80,672	81,406	89,070	90,021	90,255	90,063	95,888

FIGURE 8.4 This example shows rows of data with white space between them that exceeds the practical limit of a 1:1 ratio of text height to white space height.

White space can be intentionally manipulated to direct your readers' eyes to predominantly scan either across rows or down columns. If you wish to lead your readers to scan down columns, rather than across rows, make the white space between the columns greater than the white space between the rows. To direct readers to scan predominantly across rows, simply do the opposite.

If you don't have enough overall space on the page or screen to use white space alone to delineate the rows or columns, it's time to move on to another method.

RULES AND GRIDS

Rules and grids can be used to 1) delineate columns and rows, 2) group sections of data, and 3) highlight sections of data. Of these, the delineation of columns and rows is the least effective use of rules and grids though this use is unfortunately quite common. The problem with rules and grids is that they excessively fragment the data and introduce clutter. As you scan across rows or down columns of data that are broken up by lines, the lines distract your eyes, promoting a strong perception of individual cells through the Gestalt principle of enclosure rather than a seamless flow of information. Examples make this argument better than words. The following examples start with strong grids and gradually decrease the use of grids and rules to a minimum. Judge for yourself which table is easiest to read.

Product	Jan	Feb	Mar	Apr	May	Jun
Product 01	93,993	84,773	88,833	95,838	93,874	83,994
Product 02	87,413	78,839	82,615	89,129	87,303	78,114
Product 03	90,036	81,204	85,093	91,803	89,922	80,458
Product 04	92,737	83,640	87,646	94,557	92,620	82,872
Product 05	83,733	75,520	79,137	85,377	83,627	74,826
Total	447,913	403,976	423,323	456,705	447,346	400,264

FIGURE 8.5 This table uses a thick grid to delineate columns and rows and to enclose the entire table.

Product	Jan	Feb	Mar	Apr	May	Jun
Product 01	93,993	84,773	88,833	95,838	93,874	83,994
Product 02	87,413	78,839	82,615	89,129	87,303	78,114
Product 03	90,036	81,204	85,093	91,803	89,922	80,458
Product 04	92,737	83,640	87,646	94,557	92,620	82,872
Product 05	86,245	77,785	81,511	87,938	86,136	77,071
Total	450,425	406,241	425,697	459,266	449,854	402,508

FIGURE 8.6 In contrast to the table in *Figure 8.5*, thin horizontal rules now delineate rows in the body of the table.

Product	Jan	Feb	Mar	Apr	May	Jun
Product 01	93,993	84,773	88,833	95,838	93,874	83,994
Product 02	87,413	78,839	82,615	89,129	87,303	78,114
Product 03	90,036	81,204	85,093	91,803	89,922	80,458
Product 04	92,737	83,640	87,646	94,557	92,620	82,872
Product 05	83,733	75,520	79,137	85,377	83,627	74,826
Total	447,913	403,976	423,323	456,705	447,346	400,264

FIGURE 8.7 In contrast to the table in *Figure 8.6*, all vertical rules have now been thinned.

Product	Jan	Feb	Mar	Apr	May	Jun
Product 01	93,993	84,773	88,833	95,838	93,874	83,994
Product 02	87,413	78,839	82,615	89,129	87,303	78,114
Product 03	90,036	81,204	85,093	91,803	89,922	80,458
Product 04	92,737	83,640	87,646	94,557	92,620	82,872
Product 05	86,245	77,785	81,511	87,938	86,136	77,071
Total	450,425	406,241	425,697	459,266	449,854	402,508

FIGURE 8.8 In contrast to the table in *Figures 8.7*, all of the rules have now been lightened.

Product	Jan	Feb	Mar	Apr	May	Jun
Product 01	93,993	84,773	88,833	95,838	93,874	83,994
Product 02	87,413	78,839	82,615	89,129	87,303	78,114
Product 03	90,036	81,204	85,093	91,803	89,922	80,458
Product 04	92,737	83,640	87,646	94,557	92,620	82,872
Product 05	83,733	75,520	79,137	85,377	83,627	74,826
Total	447,913	403,976	423,323	456,705	447,346	400,264

FIGURE 8.9 In contrast to the table in *Figures 8.8*, all vertical rules have now been removed, along with the border around the entire table.

Product	Jan	Feb	Mar	Apr	May	Jun
Product 01	93,993	84,773	88,833	95,838	93,874	83,994
Product 02	87,413	78,839	82,615	89,129	87,303	78,114
Product 03	90,036	81,204	85,093	91,803	89,922	80,458
Product 04	92,737	83,640	87,646	94,557	92,620	82,872
Product 05	83,733	75,520	79,137	85,377	83,627	74,826
Total	447,913	403,976	423,323	456,705	447,346	400,264

FIGURE 8.10 In contrast to the table in *Figure 8.9*, only the rule beneath the headers and the rule above the footer row of totals now remain, and even these have been further thinned and lightened.

I think you will agree that the last of these tables is the easiest to read. It has entirely abandoned the use of grids to delineate rows and columns, and rules have only been used to separate the headers and footers from the body of the table. Nothing more is needed. Anything more would reduce the table's effectiveness.

Even using rules to form a boundary around the entire table is a waste of ink except when you must place other objects (such as additional tables, graphs, or text) on the page or screen close to the table, When you can't surround the table with sufficient white space, a light boundary around the table may be useful to separate it from the objects around it, enabling the reader to focus on it without distraction from surrounding content.

Despite the uselessness of grids and the limited usefulness of rules to delineate columns and rows in tables, grids and rules are sometimes useful for grouping or highlighting sections of data. Consider the following example, in which a column of totals has been included at the right, and additional white space has been used to group the totals as distinct from the other columns.

Product	Jan	Feb	Mar	Apr	May	Jun	Total
Product 01	93,993	84,773	88,833	95,838	93,874	83,994	541,305
Product 02	87,413	78,839	82,615	89,129	87,303	78,114	503,414
Product 03	90,036	81,204	85,093	91,803	89,922	80,458	518,516
Product 04	92,737	83,640	87,646	94,557	92,620	82,872	534,072
Product 05	83,733	75,520	79,137	85,377	83,627	74,826	482,220
Total	447,913	403,976	423,323	456,705	447,346	400,264	2,579,526

FIGURE 8.11 This example uses white space to make one column, in this case the column of totals at the right, stand out as different from the other columns.

Although using extra white space to highlight the Total column is effective, you could also use a light rule to set this column apart if space is limited:

Product	Jan	Feb	Mar	Apr	May	Jun	Total
Product 01	93,993	84,773	88,833	95,838	93,874	83,994	541,305
Product 02	87,413	78,839	82,615	89,129	87,303	78,114	503,414
Product 03	90,036	81,204	85,093	91,803	89,922	80,458	518,516
Product 04	92,737	83,640	87,646	94,557	92,620	82,872	534,072
Product 05	83,733	75,520	79,137	85,377	83,627	74,826	482,220
Total	447,913	403,976	423,323	456,705	447,346	400,264	2,579,526

FIGURE 8.12 This example uses a vertical rule to distinguish one column from the others, in this case the column of totals at the right.

Rather than the vertical rule above, we could have used four rules to form a border around the entire Total column. This would work, but it is a good practice to use the minimum ink necessary to do the job. In the following example, the objective is to highlight the March column as particularly important, which cannot be done with vertical rules alone, so the combination of vertical and horizontal rules has been used to form a border around it.

Product	Jan	Feb	Mar	Apr	May	Jun	Total
Product 01	93,993	84,773	88,833	95,838	93,874	83,994	541,305
Product 02	87,413	78,839	82,615	89,129	87,303	78,114	503,414
Product 03	90,036	81,204	85,093	91,803	89,922	80,458	518,516
Product 04	92,737	83,640	87,646	94,557	92,620	82,872	534,072
Product 05	83,733	75,520	79,137	85,377	83,627	74,826	482,220
Total	447,913	403,976	423,323	456,705	447,346	400,264	2,579,526

FIGURE 8.13 This example uses a border to highlight a particular set of data.

When you use rules, be sure to subdue them visually in relation to the data by keeping the lines thin and light. I've found that shades of gray often work best and can be relied on to remain subtle in comparison to data rendered in black, even when the table is photocopied in black and white. However, if a section of the table needs to shout, then thicker, darker lines would be appropriate.

FILL COLOR

When white space alone can't be used effectively to delineate columns and rows in tables, fill colors usually work better than grids and rules. When subtly designed, fill colors are less distracting to the eye as it scans across them. They are limited, however, to one direction; they can delineate columns or rows but not both simultaneously. Here's an example of how a light shade of gray on alternating rows aids the eye in scanning across long rows of data.

Product	Jan	Feb	Mar	Apr	May	Jun	Jul	Aug	Sep	Oct	Nov	Dec
Product 01	93,993	84,773	88,833	95,838	93,874	83,994	84,759	92,738	93,728	93,972	93,772	99,837
Product 02	87,413	78,839	82,615	89,129	87,303	78,114	78,826	86,246	87,167	87,394	87,208	92,848
Product 03	90,036	81,204	85,093	91,803	89,922	80,458	81,191	88,834	89,782	90,016	89,824	95,634
Product 04	92,737	83,640	87,646	94,557	92,620	82,872	83,626	91,499	92,476	92,716	92,519	98,503
Product 05	86,245	77,785	81,511	87,938	86,136	77,071	77,773	85,094	86,002	86,226	86,043	91,608
Product 06	88,833	80,119	83,956	90,576	88,720	79,383	80,106	87,647	88,582	88,813	88,624	94,356
Product 07	82,614	74,511	78,079	84,236	82,510	73,826	74,498	81,511	82,382	82,596	82,420	87,751
Product 08	85,093	76,746	80,421	86,763	84,985	76,041	76,733	83,957	84,853	85,074	84,893	90,384
Product 09	87,646	79,048	82,834	89,366	87,535	78,322	79,035	86,475	87,399	87,626	87,440	93,095
Product 10	90,275	81,420	85,319	92,047	90,161	80,672	81,406	89,070	90,021	90,255	90,063	95,888

FIGURE 8.14 This example uses a light gray fill color to assist horizontal scanning.

Using fill color behind every other row or every few rows in this manner to assist the eyes in scanning across rows is sometimes called *zebra striping*. As you can see, it doesn't take much. This shade of gray is very light yet visible enough to assist our eyes in scanning across a single row without confusing it with the rows above or below. The shading is light enough that vertical scanning down a column is not significantly disrupted. Notice in the next example that when background fill colors are too dark, they suffer from three problems: 1) the text is harder to read because of insufficient contrast between the background color and the text; 2) it looks like every other row has been highlighted, which is not intended; and 3) scanning down a column of numbers is complicated by the alternating light and dark background colors.

Product	Jan	Feb	Mar	Apr	May	Jun	Jul	Aug	Sep	Oct	Nov	Dec
Product 01	93,993	84,773	88,833	95,838	93,874	83,994	84,759	92,738	93,728	93,972	93,772	99,837
Product 02	87,413	78,839	82,615	89,129	87,303	78,114	78,826	86,246	87,167	87,394	87,208	92,848
Product 03	90,036	81,204	85,093	91,803	89,922	80,458	81,191	88,834	89,782	90,016	89,824	95,634
Product 04	92,737	83,640	87,646	94,557	92,620	82,872	83,626	91,499	92,476	92,716	92,519	98,503
Product 05	86,245	77,785	81,511	87,938	86,136	77,071	77,773	85,094	86,002	86,226	86,043	91,608
Product 06	88,833	80,119	83,956	90,576	88,720	79,383	80,106	87,647	88,582	88,813	88,624	94,356
Product 07	82,614	74,511	78,079	84,236	82,510	73,826	74,498	81,511	82,382	82,596	82,420	87,751
Product 08	85,093	76,746	80,421	86,763	84,985	76,041	76,733	83,957	84,853	85,074	84,893	90,384
Product 09	87,646	79,048	82,834	89,366	87,535	78,322	79,035	86,475	87,399	87,626	87,440	93,095
Product 10	90,275	81,420	85,319	92,047	90,161	80,672	81,406	89,070	90,021	90,255	90,063	95,888

FIGURE 8.15 The background fill colors in this example are too dark.

Fill colors can also be used to group and highlight sections of data. In the example below, the intention is to simply delineate the headers and footers as distinct from the rest, not to highlight them as more important:

Product	Jan	Feb	Mar	Apr	May	Jun	Total
Product 01	93,993	84,773	88,833	95,838	93,874	83,994	541,305
Product 02	87,413	78,839	82,615	89,129	87,303	78,114	503,414
Product 03	90,036	81,204	85,093	91,803	89,922	80,458	518,516
Product 04	92,737	83,640	87,646	94,557	92,620	82,872	534,072
Product 05	83,733	75,520	79,137	85,377	83,627	74,826	482,220
Total	447,913	403,976	423,323	456,705	447,346	400,264	2,579,526

FIGURE 8.16 This example uses a fill color to group particular data.

Because I didn't want to suggest that the names of the months or the totals are more important than the other data, I've used a light fill color.

In the example below, fill color has been used to highlight two cells of data:

Product	Jan	Feb	Mar	Apr	May	Jun	Total
Product 01	93,993	84,773	88,833	95,838	93,874	83,994	541,305
Product 02	87,413	78,839	82,615	89,129	87,303	78,114	503,414
Product 03	90,036	81,204	85,093	91,803	89,922	80,458	518,516
Product 04	92,737	83,640	87,646	94,557	92,620	82,872	534,072
Product 05	83,733	75,520	79,137	85,377	83,627	74,826	482,220
Total	447,913	403,976	423,323	456,705	447,346	400,264	2,579,526

FIGURE 8.17 This example uses fill color in the background to highlight particular data.

When your purpose is to highlight, the fill color should be more noticeable. Don't go overboard, though, unless you want your readers to ignore everything else.

Arranging Data

Our focus in this section is on how we can best arrange information in a table to tell its story. Should the categorical items be displayed vertically in a column or horizontally across a row? Should information be separated into subgroups with breaks in between? Should the information be sorted in a particular order?

COLUMNS OR ROWS

Categorical items can be freely arranged across columns and down rows. Some arrangements work better than others, however. When you begin to construct a table, you have already identified one or more categories that are required to communicate your message. You must now arrange the sets of categorical items across the columns or down the rows to best tell the story. How do you determine what goes where? What questions should you ask yourself about the information to make that determination? Take a minute right now to put yourself in this position, using a real or hypothetical data set. Focus on what you need to know about each category to decide whether its items should be arranged across the columns or down the rows.

. . . .

Here are the questions that I ask myself about each category:

- How many items does it contain?
- What is the maximum number of characters in any one of its items?
- Will the items display a time-series relationship or a ranking relationship?

If the answer to the question "How many items does it contain?" is "only a few," then you have the option of arranging them in a row in separate columns. If there are more than a few, you have no choice but to arrange them in a single column because of the limited width of a standard page or screen. For instance, because there are only four sales regions in the following table, they can be arranged along a row, one per column.

Product	Regions			
	North	East	South	West
Product 01	94	152	174	87
Product 02	122	198	226	113
Product 03	101	164	188	94
Product 04	142	230	263	131
Product 05	132	214	244	122
Product 06	174	282	323	161
Product 07	401	648	742	371
Product 08	281	454	519	260
Product 09	112	182	208	104
Product 10	584	944	1,081	540
Product 11	543	878	1,005	502
Product 12	163	263	301	151
Product 13	489	790	904	452
Product 14	327	529	606	303
Product 15	295	476	545	273
Total	3,960	6,403	7,330	3,665

FIGURE 8.18 This is an example of arranging categorical items, in this case sales regions, as separate columns.

If you had 15 sales regions in addition to 15 products, however, you probably wouldn't have enough space to display either in separate columns, so you might need to display both sets of categorical items down the rows with one column for products and another for regions, perhaps as follows:

Region	Product	Units Sold
Region 01	Product 01	152
	Product 02	198
	Product 03	164
	Product 04	230
	Product 05	214
	Product 06	282
	Product 07	648
	Product 08	454
	Product 09	182
	Product 10	944
	Product 11	878
	Product 12	263
	Product 13	790
	Product 14	529
	Product 15	476
Region 02	Product 01	443
	Product 02	133
	Product 03	399

FIGURE 8.19 This is an example of arranging all items in a particular category in a single column, because each set has too many items to arrange across multiple columns.

If you compare the last two examples, you might notice that by arranging both sets of categorical items side by side down the rows (*Figure 8.19*) rather than with one set down the rows and the other across the columns (*Figure 8.18*), you are forced to give up something that could be seen more readily in the bidirectional arrangement of the data. Can you see it? With the unidirectional arrangement of the data in *Figure 8.19*, you can easily compare the sales of different products within a given region, but it is more difficult to compare the sales of a product in different regions because those values aren't close together. With the bidirectional arrangement in *Figure 8.18*, the sales for the various regions are close to one another, making comparisons easier. This is worth noting. Bidirectional tables arrange more of the quantitative values closer together than unidirectional tables, offering a distinct advantage when comparisons need to be made. For this reason, you should almost always arrange the categorical items bidirectionally when you can fit one or more sets of categorical items across multiple columns.

The next question to consider is "What is the maximum number of characters in any one of the items?" If some of the items contain a large number of characters, this tells you that if you arrange them across the columns, the columns will have to be wide. Even if you have a small number of items, you may not have the horizontal space required to arrange them in wide columns. If you arrange the items down the rows in a single column, however, you only need one wide column, which saves considerable horizontal space.

Next, you should ask the question, "Will the categorical items display either a time-series or a ranking relationship?" Why does it matter? Some sequential relationships (both time and rankings have a proper sequence) are more easily understood when they are arranged in a particular way, either horizontally from left to right or vertically from top to bottom. If your table contains time-series data (years, quarters, months, etc.), what would usually work best, a left-to-right or top-to-bottom arrangement? It is more natural to think of time as moving from left to right, rather than down from top to bottom. For this reason, whenever space permits, time-series items should be arranged across the columns, as in the following example.

Region	2010				2011	
	Q1	Q2	Q3	Q4	Q1	Q2
North	393	473	539	639	439	538
East	326	393	447	530	364	447
South	401	483	550	652	448	549
West	538	647	737	874	601	736
Total	1,658	1,996	2,274	2,696	1,852	2,270

FIGURE 8.20 This example shows the preferred arrangement of time-series data across the columns from left to right.

A vertical arrangement would not work as well, illustrated below.

Year	Qtr	Region				Total
		North	East	South	West	
2010	1	393	326	401	538	1,658
	2	473	393	483	647	1,996
	3	539	447	550	737	2,273
	4	639	530	652	874	2,695
2011	1	439	364	448	601	1,852
	2	538	447	549	736	2,270

FIGURE 8.21 This example shows an awkward arrangement of time-series data.

When you display a ranking relationship, such as your top 10 products, it works best to arrange the items vertically, from top to bottom, as in the following example.

Rank	Product	Sales (U.S. $)
1	Product J	1,939,993
2	Product E	1,784,794
3	Product G	1,642,010
4	Product A	1,510,649
5	Product D	1,389,797
6	Product C	1,278,614
7	Product B	1,176,324
8	Product H	1,082,219
9	Product F	995,641
10	Product I	915,990

FIGURE 8.22 This example shows a vertical arrangement of ranked data.

Even if you had room to arrange the 10 products horizontally across the columns, you generally wouldn't want to because rankings look more natural when arranged from top to bottom.

GROUPS AND BREAKS

It is often appropriate, and perhaps even necessary, to break sets of data into smaller groups. Sometimes we do so to direct our readers to examine particular groups of data in isolation from the rest. Sometimes we do so because smaller chunks of data are easier to handle. Sometimes we are forced to break data into groups simply because we can't fit everything horizontally across the page or screen. Sometimes we create group breaks so we can display summary values for them, such as sums and averages.

Whenever you break the data into smaller chunks and arrange them vertically, do so logically, based on categories. For instance, you could start a new group of information for each year or one for each department in your organization. You might need to break the data into even smaller chunks, basing the groups on a combination of categories, such as country and sales region, as illustrated in the following example.

Country: USA   Region: North

Product	Jan	Feb	Mar	Apr	May	Jun	Jul	Aug	Sep	Oct	Nov	Dec
Product 01	93,993	84,773	88,833	95,838	93,874	83,994	84,759	92,738	93,728	93,972	93,772	99,837
Product 02	87,413	78,839	82,615	89,129	87,303	78,114	78,826	86,246	87,167	87,394	87,208	92,848
Product 03	90,036	81,204	85,093	91,803	89,922	80,458	81,191	88,834	89,782	90,016	89,824	95,634
Product 04	92,737	83,640	87,646	94,557	92,620	82,872	83,626	91,499	92,476	92,716	92,519	98,503
Product 05	86,245	77,785	81,511	87,938	86,136	77,071	77,773	85,094	86,002	86,226	86,043	91,608
Product 06	88,833	80,119	83,956	90,576	88,720	79,383	80,106	87,647	88,582	88,813	88,624	94,356
Product 07	82,614	74,511	78,079	84,236	82,510	73,826	74,498	81,511	82,382	82,596	82,420	87,751
Product 08	85,093	76,746	80,421	86,763	84,985	76,041	76,733	83,957	84,853	85,074	84,893	90,384
Product 09	87,646	79,048	82,834	89,366	87,535	78,322	79,035	86,475	87,399	87,626	87,440	93,095
Product 10	90,275	81,420	85,319	92,047	90,161	80,672	81,406	89,070	90,021	90,255	90,063	95,888
Total	$884,886	$798,085	$836,307	$902,255	$883,765	$790,751	$797,953	$873,070	$882,391	$884,688	$882,805	$939,903

Country: USA   Region: East

Product	Jan	Feb	Mar	Apr	May	Jun	Jul	Aug	Sep	Oct	Nov	Dec
Product 01	93,993	84,773	88,833	95,838	93,874	83,994	84,759	92,738	93,728	93,972	93,772	99,837
Product 02	87,413	78,839	82,615	89,129	87,303	78,114	78,826	86,246	87,167	87,394	87,208	92,848
Product 03	90,036	81,204	85,093	91,803	89,922	80,458	81,191	88,834	89,782	90,016	89,824	95,634

FIGURE 8.23 This example shows a grouping of data based on a combination of categories, in this case country and region.

Notice that if you want to sum sales by region, it would be awkward to incorporate them into the table without breaking the data into regional groups with group footers. Notice also that the group headers (e.g., Country: USA, Region: North) interrupt vertical scanning of the information. This isn't a problem in this case because the interruption reinforces the fact that a new group has begun and causes readers to briefly focus on the header information.

Whatever your reason for breaking the data into groups, keep in mind the following design practices:

- Use vertical white space between the groups but only enough to make the break noticeable. Excessive white space does nothing but spread the data farther apart and reduce the amount of information on the page or screen, which is rarely useful.
- Repeat the column headers at the beginning of each new group. This will help your readers keep track of the content in each column.
- Don't vary the structure of the table from group to group. For instance, each group should contain the same columns, in the same order, with the same widths; otherwise, your readers will be forced to relearn the structure at the beginning of each new group.
- When you group data based on multiple categories (e.g., country and region), position the group headers on the same row, in order, from left to right, unless there isn't enough horizontal space to do so. This arrangement saves vertical space and clearly represents the hierarchical relationship between the categorical sets. *Figure 8.23*, on the previous page illustrates this arrangement.
- When you want your readers to examine each group of data in isolation from the rest, start each group on a new page.

These design practices are not arbitrary. Each of them is effective because it corresponds to the way that visual perception works.

COLUMN SEQUENCE

The primary considerations when you are determining the best order in which to arrange the columns of a table are the following:

- Each set of categorical items that is arranged down the rows of a single column should be placed to the left of the quantitative values that are associated with it.
- If there is a hierarchical relationship between categories (e.g., between product families and products), they should be sequenced from left to right to reflect that order.
- Quantitative values that are derived from another set of quantitative values using a calculation should generally be placed in a column just to the right of the column from which they were derived.

Let's look at an example of this last design practice before continuing the list.

Product	Units Sold	Actual Revenue	% of Total	Fcst Revenue	% of Fcst
Product A	938	187,600	47%	175,000	107%
Product B	1,093	114,765	28%	130,000	88%
Product C	3,882	62,112	15%	50,000	124%
Product D	873	36,666	9%	40,000	92%
Product E	72	2,088	1%	50,000	4%
Total	6,858	$403,231	100%	$445,000	91%

FIGURE 8.24 This example shows columns of quantitative values that are derived (i.e., calculated) from a source column and are therefore placed to the right of the source column.

In this example, even though the % of Total column is somewhat ambiguous because it doesn't clarify what the percentages are based on, most readers would still assume that it is the percentage of revenue for each product compared to the total revenue for all products. They would do so because this column is to the right of the Actual Revenue column. Even if its header were clearer (e.g., % of Total Actual Revenue), confusion would result if it were placed to the right of the Units Sold column or the Fcst Revenue column.

Now, back to our list of design practices.

- Columns containing sets of quantitative values that you want your readers to easily compare should be placed as close to one another as possible.

In the following example, last year's revenue has intentionally been placed next to this year's revenue to encourage comparisons between them.

Product	Units Sold	Last Year's Revenue	This Year's Revenue	Fcst Revenue	Planned Revenue
Product A	938	159,497	187,600	175,000	160,000
Product B	1,093	123,007	114,765	130,000	125,000
Product C	3,882	45,384	62,112	53,000	50,000
Product D	873	41,003	36,666	38,000	40,000
Product E	72	2,485	2,088	4,000	5,000
Total	6,858	$371,376	$403,231	$400,000	$380,000

FIGURE 8.25 This example places particular columns next to one another to encourage comparisons.

There is one more design practice that deserves a place on the list, but I want you to work on your own a little for this one. Imagine that you are designing a table that will display units sold for a set of five products (i.e., products A through E) and a set of four sales channels (i.e., Direct, Reseller, Distributor, and Original Equipment Manufacturer [OEM]). You will have three columns: one for units sold, one for products, and one for sales channels. If you want to help your readers easily compare sales through the various channels for each product, in what order would you place the columns?

· · · · · · ·

To achieve this objective, you would place the sales channel column to the right of the product column. Here's an example of how it might look, with an additional column of percentages thrown in to enhance the comparisons.

Product	Channel	Units Sold	% of Total
Product A	Direct	8,384	26.5%
	Reseller	7,384	23.4%
	Distributor	10,838	34.3%
	OEM	4,993	15.8%
Product B	Direct	5,939	23.5%
	Reseller	7,366	29.1%
	Distributor	8,364	33.0%
	OEM	3,645	14.4%

FIGURE 8.26 This example arranges columns containing categories in a particular sequence to support comparisons that are important to the message.

Here's a statement of this design practice:

- To enable easy comparisons between individual items in a particular category, either arrange them across multiple columns if space allows, or in a single column to the right of the other columns of categorical data.

That's quite a mouthful, but I hope the example we worked through helps you make sense of it. In the example, because you wanted to help your readers make comparisons among sales through the various channels for individual products, the channel column had to go to the right of the product column. Placing the product column to the right of the channel column would not have served this purpose.

DATA SEQUENCE

The common term for sequencing data in a table is *sorting*. Numbers and dates both have a natural order that is meaningful. When you need to sequence numbers, their quantitative order, either ascending or descending, is the only useful way to sequence them. The same is true regarding the chronological order of dates and times. Other data, such as the names of things (e.g., products, customers, and countries) and other types of identifiers (e.g., purchase order numbers) have a conventional order based on alphabetical sequence, which is useful for look-up purposes but isn't meaningful otherwise. Alphabetical order is useful in tables for items that have no natural order, such as the names of your customers because this arrangement makes it easy for your readers to find any customers that they're looking for by scanning alphabetically through the list.

However, when a set of categorical items has a natural order, sorting them in alphabetical order would be unnatural and confusing. For instance, if your company does business in five countries—the United States, England, France, Australia, and Germany—and 80% of your business is done is the United States, and the percentage of business conducted in the other countries declines in the order in which they are listed above, to list them alphabetically in tables would make no sense. The point is simple: if items have a natural order that is meaningful, sort them in that order. This will not only create a sequence that makes sense but will also place near one another those items that your readers will often want to compare.

Formatting Text

In this section we'll focus on aspects of text formatting that play a role in table design. By text, I'm referring to all alphanumeric characters, including numbers and dates.

ORIENTATION

As speakers of English, we are accustomed to reading language from left to right, arranged horizontally. Any other orientation is more difficult to read, as a quick scan of the following should confirm.

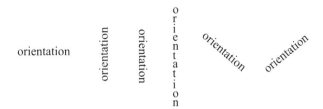

FIGURE 8.27 This figure illustrates various ways in which text can be oriented.

There is rarely a valid reason to sacrifice the legibility of text in a table by orienting it in any way other than horizontally from left to right. As we'll see in the next chapter, an alternate orientation is sometimes required in graphs, but this is seldom the case in tables.

ALIGNMENT

The most effective alignment of text is primarily a matter of convention rather than the mechanics of visual perception, but this convention is powerful, so ignoring it will lead to inefficiency, frustration, and confusion. The best practices of alignment can be summarized as follows:

- Numbers that represent quantitative values, as opposed to those that are merely identifiers (e.g., customer numbers), should always be aligned to the right.
- Text that expresses neither numbers nor dates works best when aligned to the left.
- How dates are aligned doesn't matter, but it's important to format them in a way that keeps the number of digits in each part of the date (i.e., month, day, and year) consistent.

Aligning numbers to the left or center makes them difficult to compare, as illustrated in the following example:

Sales	Sales	Sales
93,883.39	93,883.39	93,883.39
5,693,762.32	5,693,762.32	5,693,762.32
483.84	483.84	483.84
674,663.39	674,663.39	674,663.39
548.93	548.93	548.93
3,847.33	3,847.33	3,847.33
$6,467,189.20	$6,467,189.20	$6,467,189.20

FIGURE 8.28 These examples show the difficulty created when numbers are not aligned to the right.

This preference holds true for all units of measure, including money and percentages. You might be concerned that this preference for aligning numbers to the right would not hold true if the values in a particular column contained decimal digits of varying number, in which case the decimal points would not be aligned. It is best in such cases is to align both the decimal point and the final digit to the right. This can be accomplished by expressing each value using the same number of decimal digits, even when they are zeroes, as in the table on the right in the following example. Notice how the ragged right edge that's

created by an inconsistent number of decimal digits gives the table on the left a
sloppy appearance and makes comparisons more difficult.

Rate	Rate
3.5%	3.500%
12.675%	12.675%
5.%	5.000%
13.25%	13.250%
2.75%	2.750%
13.125%	13.125%
8.383%	8.383%

FIGURE 8.29 This example
compares the right alignment of
numbers to the decimal point with
an inconsistent number of decimal
digits on the left and a consistent
number on the right, which works
best.

Text that expresses neither numbers nor dates works best when aligned to the
left because we read from left to right. This includes numbers that are used as
identifiers, rather than quantitative values, as is the case with customer numbers, for example. Here are examples of various text alignments:

Product Code	Product Name	Product Code	Product Name	Product Code	Product Name
A1838	2-Door Sport	A1838	2-Door Sport	A1838	2-Door Sport
A89	4-Door Sport	A89	4-Door Sport	A89	4-Door Sport
J98488	2-Door Luxury	J98488	2-Door Luxury	J98488	2-Door Luxury
J3883	4-Door Luxury	J3883	4-Door Luxury	J3883	4-Door Luxury
K9288	2-Door Truck	K9288	2-Door Truck	K9288	2-Door Truck
K38733	4-Door Truck	K38733	4-Door Truck	K38733	4-Door Truck

FIGURE 8.30 These examples show
various text alignments, including
the preferable left alignment
highlighted in gray.

Don't be tempted by the aesthetic appeal of centered text in the columns of
tables. The ragged left edge of the text makes scanning less efficient than the
consistent leading edge of left-aligned text. I find that one exception to the
practice of left alignment works well for columns of text, however: when the
text entries in a column each consist of the same number of characters and the
column header consists of several more characters than the text entries. Here are
two examples using columns with single character text entries, which display
left and center alignment:

Cust Code	Preferred?	Cust Code	Preferred?
193847394	Y	193847394	Y
109388484	N	109388484	N
187466463	N	187466463	N
174563553	N	174563553	N
175357736	Y	175357736	Y
167374565	Y	167374565	Y

FIGURE 8.31 Theses examples show
left- versus center-aligned text for
columns with lengthy headers and
single-character text entries.

Centering rather than left aligning such text entries does not produce a ragged
left edge, so reading is not impaired. Centering also has the advantage of adding
more white space to the left of the text entries, thus distinguishing them more
clearly from data in adjacent columns.

There is no strict convention for the alignment of dates. People differ in their
preferences for left, right, and center alignment, and, in truth, it doesn't matter.
What does matter, however, is that each part of the date is composed of a
consistent number of digits (e.g., 01/02/11 rather than 1/2/11 or 01/02/2011

rather than 1/2/11). When the number of digits is consistent, the parts of the dates in a column are always aligned for easy comparison. The following example illustrates some of the alternatives, with my preferences highlighted in gray.

Left	Left	Center	Center	Right	Right
12/17/02	12/17/02	12/17/02	12/17/02	12/17/02	12/17/02
1/2/03	01/02/03	1/2/03	01/02/03	1/2/03	01/02/03
1/17/03	01/17/03	1/17/03	01/17/03	1/17/03	01/17/03
2/9/03	02/09/03	2/9/03	02/09/03	2/9/03	02/09/03
10/29/03	10/29/03	10/29/03	10/29/03	10/29/03	10/29/03
12/1/03	12/01/03	12/1/03	12/01/03	12/1/03	12/01/03
1/1/03	01/01/03	1/1/03	01/01/03	1/1/03	01/01/03

FIGURE 8.32 These examples show left, center, and right alignment of dates, using two different date formats.

When we use a consistent number of digits in each date, each part of the date (month, day, and year) is aligned as well, making comparisons of a specific component (e.g., year) easy and efficient.

You may have noticed that in each of the examples above, the headers are aligned with the associated data. If entries in the column are left aligned, the column header is left aligned as well, and so on. This is intentional. This practice clearly establishes at the top of each column the nature of its alignment and it creates a clear association between the column's header and information in the body of that column. The only exception that is sometimes useful involves spanner headers. When a header is used to label multiple columns, centering it across those columns often helps to clarify the fact that it refers to all of those columns rather than just a single column. This is illustrated by the centering of the header Region in the following example:

Year	Qtr	Region				Total
		North	East	South	West	
2010	1	393	326	401	538	1,658
	2	473	393	483	647	1,996
	3	539	447	550	737	2,273
	4	639	530	652	874	2,695
2011	1	439	364	448	601	1,852
	2	538	447	549	736	2,270

FIGURE 8.33 This example shows a spanner header as an exception to the normal rules of alignment.

NUMBER AND DATE FORMAT

The format that works best for displaying numbers and dates is the one that exhibits the following characteristics:

- Includes no unnecessary information (e.g., excessive levels of precision, which we'll examine in the next section)
- Expresses information using the format that is most familiar to your readers
- Most consistently aligns information from row to row for efficient scanning

Rather than list and comment on all possible variations of number and date formats, I'll highlight only those formatting practices that communicate effectively.

Here's a list of the useful practices for formatting numbers in tables:

- Place a comma (or a space for European readers) to the left of every three whole-number digits (e.g., 1,393,033 rather than 1393033).
- Truncate the display of whole numbers by sets of three digits to the nearest thousand, million, billion, etc., whenever numeric precision can be reduced without the loss of meaningful information, and declare that you've done so in the title or header (e.g., "U.S. dollars in thousands").
- Use either parentheses or a negative sign to display negative numbers (e.g., (8,395.37) or -8,395.37), but if you use parentheses, keep the numbers that are enclosed in them right aligned with the positive numbers.

The first three practices in this list are each firmly rooted in what you now know about visual perception. The commas break the numbers up into smaller chunks that can be stored more easily in working memory. They also make it easy for the reader to determine the difference between numbers such as 10000000 and 1000000 at a glance (i.e., 10,000,000 and 1,000,000) without having to count the digits. Truncating whole numbers to remove everything below 1,000, or everything below 1,000,000, or etc., reduces the amount of information that must be read, which makes the process of reading more efficient, and reduces the amount of horizontal space required by the column. Aligning the right-most digits of positive and negative numbers, even when the negative numbers are enclosed in parentheses, keeps the numbers aligned for easy scanning down the column. Look at the difference between the two formats below:

83,743	83,743
2,339,844	2,339,844
(67,909)	(67,909)
60,036	60,036
376,003	376,003
3,974,773	3,974,773
(576,533)	(576,533)
937,764	937,764
343	343

FIGURE 8.34 This example shows two alignment techniques for negative numbers that are enclosed within parentheses, with the preferred method highlighted in gray.

Readers' eyes would be forced to jump back and forth slightly from row to row when scanning the values in the right-hand column in the example above.

Now, here's a list of useful practices for formatting dates in tables:

- Express months either as a two-digit number (e.g., 02 rather than 2 for February) or a three-character word (e.g., Feb rather than February).
- Express days using two digits (e.g., 01 rather than 1 for the first day of the month).
- Use a format that excludes portions of the date that provide more precision than necessary.

The first two practices relate to the advantages of alignment for efficient visual perception, which we covered earlier. The last relates entirely to the level of precision that is appropriate, which we'll cover next.

If your readers are distributed internationally, keep in mind that conventional date formats differ in various parts of the world. The difference that is most prone to create confusion is that in the United States we usually list the month first, then the day, but it is common in Europe to list the day first, then the month. This isn't confusing when you express months as words rather than numbers (e.g., Dec rather than 12), but when months are expressed as numbers using a format like 01-12-11, the positions of the month and day are ambiguous. When this is the case, be sure to clarify the positions of the month and day portions of the date in your header, using a method like the following:

Order Date
(mm/dd/yy)
12/17/02
01/02/03
01/17/03

FIGURE 8.35 This example includes information in the column header to clarify the date format.

NUMBER AND DATE PRECISION

The precision of a number or a date is the degree of detail that it expresses. The number 12.825 represents a quantitative precision of three decimal digits while the number 12 only displays precision to the nearest whole number. The date December 15, 2003 represents precision to the day level while the date 2003 only displays year-level precision. Selecting the appropriate level of precision for numbers and dates in tables boils down to a single design practice:

The level of precision should not exceed the level needed to serve your communication objectives and the needs of your readers.

If the purpose of the table is to support the reconciliation of financial accounts, you had better display precision down to the penny (i.e., two decimal digits). If, however, you are presenting a multi-year comparison of sales revenue to the executives of a multi-billion dollar corporation, precision to the nearest million dollars may be appropriate. Forcing them to read six more digits of precision (or eight more digits if you include cents) would waste their time. If you express numbers with less precision than is available, you should always be careful to clearly state this in the title or header (e.g., Rounded to the nearest million dollars).

We are faced with choices regarding numeric precision whenever we produce new numbers as a result of calculations. When you divide 100 by 49 using spreadsheet software, you will likely get a result like 2.040816327. For most purposes, precision to the level of nine decimals digits is excessive. Here are examples of the excessive levels of precision you would typically get using spreadsheet software to divide numbers into 100.

```
100  /   15  =  6.666666667
100  /   20  =  5.000000000
100  /   42  =  2.380952381
100  /   49  =  2.040816327
100  /   50  =  2.000000000
100  /   55  =  1.818181818
100  /   60  =  1.666666667
100  /  175  =  0.571428571
```

FIGURE 8.36 This example shows a level of precision that exceeds what is needed for most purposes.

It is probably obvious that you would rarely need nine digits of decimal precision, but how many digits do you need? The answer depends on the message you're trying to communicate. For many purposes, rounding to the nearest whole number works fine. Here are three sample levels of precision for the list of numbers in the example above.

Decimal Digits		
0	1	2
7	6.7	6.67
5	5.0	5.00
2	2.4	2.38
2	2.0	2.04
2	2.0	2.00
2	1.8	1.82
2	1.7	1.67
1	0.6	0.57

FIGURE 8.37 These examples show three different levels of precision for the same set of values.

It isn't likely that more than a single decimal digit of precision would be significant in this case. In the example below, revenue is displayed as whole dollars per region along with each region's percentage contribution to the whole. The percentages have been repeated in four separate columns to illustrate different levels of precision.

Region	Revenue	% of Total	% of Total	% of Total	% of Total
Americas	636,663,663	40%	39.8%	39.82%	39.816%
Europe	443,874,773	28%	27.8%	27.76%	27.759%
Asia	399,393,993	25%	25.0%	24.98%	24.978%
Australia	99,838,333	6%	6.2%	6.24%	6.244%
Middle East	10,399,383	1%	0.7%	0.65%	0.650%
Africa	7,939,949	1%	0.5%	0.50%	0.497%

FIGURE 8.38 This example shows four different levels of precision for the display of the same set of percentage values.

In this example, if you needed to display both a part-to-whole and a ranking relationship between the six regions using only percentages (i.e., excluding the actual revenue dollars), which level of precision should you select for the % of Total column?

. . . .

Whole percentages (i.e., no decimal digits) would not differentiate the relative contributions or rank of the Middle East and Africa regions, and two decimal digits would provide more detail than necessary, so one decimal digit would probably work best.

Appropriate date precision is simpler to determine than numeric precision. Generally, you just need to decide whether your message requires precision to the level of year, quarter, month, week, or day. Once you determine which to use, select a date format that displays nothing below that level. Forcing your readers to read entire dates (e.g., 11/27/2011) when all they need to know is the year (e.g., 2011) not only slows them down, but also uses up more horizontal space, which is often in short supply.

FONT

There are hundreds of available fonts. A thorough knowledge of typography involves years of study and practice. Fortunately, we don't need to be experts in typography when designing tables. The best practices regarding font selection for tables have three primary goals:

- The font should be as legible as possible.
- The same font should be used throughout.
- Each numeric digit (0-9) should be equal in width.

Because our design goal is effective communication, fonts should be easy to read. Extravagant font choices intended to spice up the appearance of a table will reduce communication efficiency. This is seldom, if ever, an advantage in quantitative tables.

Fonts are generally classified into two primary types: 1) *serif* and 2) *sans-serif.* A serif is an embellishment, such as a line or curve, located at the end of a letterform.

Fonts that are most legible tend to have a clean and simple design. Although serifs are embellishments, a font with serifs can still have a clean and simple appearance. In fact, many experts have argued that serif fonts are more legible because the serif creates a greater distinction between the individual characters. Whether this is true or not isn't entirely clear, even after a comprehensive review of the research. Serif and sans-serif fonts both work fine in print as long as you select one that is highly legible. Sans-serif fonts have a slight advantage in legibility when information is displayed on a computer screen, however. Despite improvements in screen resolution, the resolution of print on paper is still higher. The lesser resolution of a screen cannot produce perfectly clean and clear serifs because they are tiny. The figure below shows examples of some of the most legible serif and sans-serif fonts as well as some that are not so legible.

Fine Legibility		Poor Legibility	
Serif	Sans-Serif	Serif	Sans-Serif
Times New Roman	Arial	Academy Engraved	**Poplar Std Black**
Palatino	Verdana	*Script*	SYNCHRO LET
Georgia	Helvetica	**Playbill**	SILOM

FIGURE 8.39 These are examples of serif and sans-serif fonts with fine and poor legibility.

This list of legible fonts is not comprehensive by any means. It gives just a few examples of good ones that are readily available to most computer users. The best typefaces for text assign different widths to characters based on character shape. When characters have equal widths regardless of shape, which was standard on typefaces created for typewriters such as `Courier`, `text is harder to read`. Numbers, however, should have equal widths; otherwise, they will not line up properly in tables, as you can see by trying to compare the numbers in the right-hand table on the following page.

Arial			Big Caslon Medium	
Region	Revenue		Region	Revenue
Americas	639,453,661		Americas	639,453,661
Europe	413,874,773		Europe	413,874,773
Asia	199,393,992		Asia	199,393,992
Australia	67,802,333		Australia	67,802,333
Middle East	10,349,381		Middle East	10,349,381
Africa	7,011,159		Africa	7,011,159

FIGURE 8.40 The best typefaces for tables vary the width of text characters but keep numbers equal in width.

Using the same font throughout a table is usually the best practice. There is almost always a better way to group or highlight particular content than using a different font. The use of emphasis (e.g., boldfacing) is one strategy that is almost always preferable to mixing fonts.

EMPHASIS AND COLOR

Text may be grouped or highlighted by using emphasis and color. Emphasis, in this context, refers to boldfacing or italicizing the text. Boldfacing makes text heavier (the lines that form characters are slightly thicker than for regular text) and italics make text stand out through a contrast in orientation, specifically by slanting the characters at an angle. Don't overdo the use of italics, however, because it is slightly harder to read than regular text. Color and emphasis can both be used effectively to group and highlight data.

Summarizing Values

Tables provide an ideal means to combine detail and summary values, simultaneously displaying high- and low-level perspectives. Values can be aggregated across an entire row or down an entire column. In Chapter 2, *Simple Statistics to Get You Started*, we examined several numbers that can be used to aggregate data, such as sums and averages. Aggregates give your readers an overview in a single glance. If readers wish, they can then dig into the detailed values. This is a powerful path to understanding, moving from general to specific, macroscopic to microscopic, summary to detail. Aggregates that are most useful in tables include:

- Measures of sum (i.e., the simple addition of a set of values)
- Measures of average (primarily means and medians)
- Measures of occurrence (i.e., a count of the number of instances of a thing or event)
- Measures of distribution (primarily the minimum and maximum values to represent the range and occasionally the standard deviation)

COLUMN AND ROW SUMMARIES

Most aggregates that appear in tables are used to summarize the values down an entire column or across an entire row. A value that aggregates an entire column of values is a *column summary*, which generally appears in a separate row beneath the last row of detailed values. A value that aggregates an entire row of values is a *row summary*, which generally appears in a separate column to the right of the last column of detailed values. The following example includes both:

Product	Jan	Feb	Mar	Apr	May	Jun	Total
Product 01	93,993	84,773	88,833	95,838	93,874	83,994	**$541,305**
Product 02	87,413	78,839	82,615	89,129	87,303	78,114	**$503,414**
Product 03	90,036	81,204	85,093	91,803	89,922	80,458	**$518,516**
Product 04	92,737	83,640	87,646	94,557	92,620	82,872	**$534,072**
Product 05	86,245	77,785	81,511	87,938	86,136	77,071	**$496,687**
Product 06	88,833	80,119	83,956	90,576	88,720	79,383	**$511,587**
Product 07	82,614	74,511	78,079	84,236	82,510	73,826	**$475,776**
Product 08	85,093	76,746	80,421	86,763	84,985	76,041	**$490,049**
Product 09	87,646	79,048	82,834	89,366	87,535	78,322	**$504,751**
Product 10	90,275	81,420	85,319	92,047	90,161	80,672	**$519,893**
Total	**$884,886**	**$798,085**	**$836,307**	**$902,255**	**$883,765**	**$790,751**	**$5,096,049**

FIGURE 8.41 This is an example of a table that includes both column and row summaries.

The column and row summaries in this example have been boldfaced simply to make them stand out in the example, not to suggest that you should always highlight summary values.

Column and row summaries frequently convey critical information. Whenever you design a table, ask yourself whether column and/or row summaries would be meaningful and useful even if they weren't specifically requested. If the answer is "yes" or even "maybe," and there is space, include them. Because they are located at the edges of the table, readers can easily ignore them if they wish.

GROUP SUMMARIES

Group summaries are similar to column and row summaries, differing only in that they aggregate meaningful subsets of values rather than the entire set of values throughout a column or row. When you group and break rows of data into subsets based on one or more categories (e.g., sales regions), it is easy to include summary values for each of those groups, as illustrated below:

Region: North

Product	Jan	Feb	Mar	Apr	May	Jun
Product 01	93,993	84,773	88,833	95,838	93,874	83,994
Product 02	87,413	78,839	82,615	89,129	87,303	78,114
Product 03	90,036	81,204	85,093	91,803	89,922	80,458
Product 04	92,737	83,640	87,646	94,557	92,620	82,872
Product 05	86,245	77,785	81,511	87,938	86,136	77,071
Product 06	88,833	80,119	83,956	90,576	88,720	79,383
Product 07	82,614	74,511	78,079	84,236	82,510	73,826
Product 08	85,093	76,746	80,421	86,763	84,985	76,041
Product 09	87,646	79,048	82,834	89,366	87,535	78,322
Product 10	90,275	81,420	85,319	92,047	90,161	80,672
Total	**$884,886**	**$798,085**	**$836,307**	**$902,255**	**$883,765**	**$790,751**

FIGURE 8.42 This is an example of group summaries, in this case per region.

Region: East

Product	Jan	Feb	Mar	Apr	May	Jun
Product 01	93,993	84,773	88,833	95,838	93,874	83,994
Product 02	87,413	78,839	82,615	89,129	87,303	78,114
Product 03	90,036	81,204	85,093	91,803	89,922	80,458

Summaries of individual values across multiple columns in a single row are placed in a column of their own, generally to the right of the values they summarize. Here's a typical example:

Product	Jan	Feb	Mar	Q1 Total	Apr	May	Jun	Q2 Total
Product 01	93,993	84,773	88,833	267,599	95,838	93,874	83,994	273,706
Product 02	87,413	78,839	82,615	248,867	89,129	87,303	78,114	254,547
Product 03	90,036	81,204	85,093	256,333	91,803	89,922	80,458	262,183
Product 04	92,737	83,640	87,646	264,023	94,557	92,620	82,872	270,048
Product 05	86,245	77,785	81,511	245,541	87,938	86,136	77,071	251,145
Product 06	88,833	80,119	83,956	252,908	90,576	88,720	79,383	258,679
Product 07	82,614	74,511	78,079	235,204	84,236	82,510	73,826	240,572
Product 08	85,093	76,746	80,421	242,260	86,763	84,985	76,041	247,789
Product 09	87,646	79,048	82,834	249,528	89,366	87,535	78,322	255,223
Product 10	90,275	81,420	85,319	257,014	92,047	90,161	80,672	262,879
Total	$884,886	$798,085	$836,307	$2,519,278	$902,255	$883,765	$790,751	$2,576,771

FIGURE 8.43 This is an example of group column summaries, in this case summing revenue by quarter.

When intermingling group summary columns with detail columns, take extra care to make a clear distinction between the two. You certainly don't have to make them as distinct as I have using gray background fills in the example above (and you shouldn't do this unless you want them to stand out as more important than the detail columns), but you should make sure that your readers notice that the summaries and details are different. In the example below, where no visual distinction is made between the detail and summary columns, readers might mistakenly perceive the summaries as just another column of monthly values:

Product	Jan	Feb	Mar	Q1 Mo Avg	Apr	May	Jun	Q2 Mo Avg
Product 01	93,993	84,773	88,833	89,200	95,838	93,874	83,994	91,235
Product 02	87,413	78,839	82,615	82,956	89,129	87,303	78,114	84,849
Product 03	90,036	81,204	85,093	85,444	91,803	89,922	80,458	87,394
Product 04	92,737	83,640	87,646	88,008	94,557	92,620	82,872	90,016
Product 05	86,245	77,785	81,511	81,847	87,938	86,136	77,071	83,715
Product 06	88,833	80,119	83,956	84,303	90,576	88,720	79,383	86,226
Product 07	82,614	74,511	78,079	78,401	84,236	82,510	73,826	80,191
Product 08	85,093	76,746	80,421	80,753	86,763	84,985	76,041	82,596
Product 09	87,646	79,048	82,834	83,176	89,366	87,535	78,322	85,074
Product 10	90,275	81,420	85,319	85,671	92,047	90,161	80,672	87,626
Total	$884,886	$798,085	$836,307	$839,759	$902,255	$883,765	$790,751	$858,924

FIGURE 8.44 This is an example of group column summaries that are not visually distinct from the columns that contain detailed data.

This tendency to confuse summary columns with detail columns becomes more pronounced as the reader's eyes move farther down the table, away from the top where the only clue to the column's identity is in the headers. You can create the minimum necessary visual distinction through the use of any one of the available visual attributes that we've already examined, including something as simple as extra white space separating the summary columns from the detail columns. Just be careful to keep the distinction subtle unless the summary columns deserve greater attention than the detail columns.

HEADERS VERSUS FOOTERS

Values that summarize rows don't necessarily have to appear below the rows that they summarize. Although it is common practice to place them in table or group footers, there is, at times, an advantage to placing them above the rows that they summarize. When might it be beneficial to place summary values before the information that they summarize rather than after it? Take a moment to imagine tables that you commonly use to see whether you can think of circumstances when the placement of summary values in headers would offer an advantage to your readers.

· · · · · · ·

When the summary values are more important to your message or to your readers than the details, and placing them below the details would make them harder and less efficient to find, it often makes sense to place them near the group headers even though this is less conventional. Here are two examples of how this can be arranged:

Region: North	$1,568,586	$1,414,719	$1,482,474	$1,599,376	$1,566,600	$1,401,719
Salesperson	Jan	Feb	Mar	Apr	May	Jun
Abrams, S	93,993	84,773	88,833	95,838	93,874	83,994
Benson, J	87,413	78,839	82,615	89,129	87,303	78,114
James, R	86,245	77,785	81,511	87,938	86,136	77,071
Wilson, O	86,704	78,199	81,944	88,406	86,594	77,481
Yao, J	89,305	80,545	84,403	91,058	89,192	79,805

FIGURE 8.45 This illustrates the placement of column summary values in a header, in this case in the same row as a group subdivision (i.e., the north region).

Region: North						
Total	$1,568,586	$1,414,719	$1,482,474	$1,599,376	$1,566,600	$1,401,719
Salesperson	Jan	Feb	Mar	Apr	May	Jun
Abrams, S	93,993	84,773	88,833	95,838	93,874	83,994
Benson, J	87,413	78,839	82,615	89,129	87,303	78,114
James, R	86,245	77,785	81,511	87,938	86,136	77,071
Wilson, O	86,704	78,199	81,944	88,406	86,594	77,481
Yao, J	89,305	80,545	84,403	91,058	89,192	79,805

FIGURE 8.46 This is an example of placing column summary values in a header, in this case in their own row beneath the group value because there isn't enough room to place them on the same row as the group label.

This same practice works just as well with grand totals. You might earn your readers' gratitude if you eliminate the hassle of having to flip to the last page of the report to get the big picture that they need most. If you're concerned that they'll be thrown off by not finding summary values where they conventionally appear, after the details, you may put the summaries there as well. Summary values are generally so useful that the little loss of space used to place them in both the headers and the footers is a small price to pay for the advantage of greater availability.

Page Information

Because of working memory constraints, effective table design requires that certain information be repeated on each new page. Otherwise, your readers may lose track of information that's needed to interpret the table as they move from page to page. Two types of information should be repeated:

- Column headers
- Row headers

REPEATING COLUMN HEADERS

When tables extend down multiple pages, columns are no longer labeled after the first page unless you repeat the column headers on each. The space required to repeat the column headers is insignificant compared to the benefit of improved ease of use.

REPEATING ROW HEADERS

I've found that in actual practice, even those of us who are careful to repeat column headers on each page seldom think to repeat the row headers as well, yet the problem is the same. In the example below, state names (Alabama, Arkansas, etc.) appear in the State column.

State	Order Date	Sales Volume	Sales Revenue
Alabama	05/01/03	432	215,568
	05/02/03	534	266,466
	05/03/03	466	232,534
	05/04/03	354	176,646
	05/05/03	456	227,544
	05/08/03	553	275,947
	05/09/03	465	232,035
	05/12/03	589	293,911
	05/15/03	501	249,999
	05/16/03	556	277,444
	05/17/03	623	310,877
	05/19/03	563	280,937
	05/22/03	675	336,825
	05/23/03	702	350,298
	05/24/03	658	328,342
	05/26/03	798	398,202
	05/29/03	801	399,699
	05/30/03	735	366,765
	05/31/03	802	400,198
Arkansas	05/01/03	201	100,299
	05/02/03	247	123,253
	05/03/03	245	122,255
	05/04/03	277	138,223
	05/05/03	203	101,297

FIGURE 8.47 This example shows a typical table that groups data, in this case by state.

Because all the rows pertaining to each state are grouped together, it isn't necessary to repeat the name of the state on each row. In fact, to do so would be inefficient, causing readers to scan, on each new row, information that they already know. Also, the infrequent presence of text in the state column alerts readers through an obvious visual cue when information for a new state has begun. However, if information for the state of Alabama continues for five pages, by the time that you reached the third page, you might have forgotten which state you were examining, especially if your attention was pulled away, even briefly. You would see something like the next example:

State	Order Date	Sales Volume	Sales Revenue
	05/04/03	354	176,646
	05/05/03	456	227,544
	05/06/03	556	277,444
	05/07/03	598	298,402
	05/08/03	553	275,947
	05/09/03	465	232,035
	05/10/03	434	216,566
	05/11/03	676	337,324
	05/12/03	589	293,911
	05/13/03	688	343,312
	05/14/03	701	349,799
	05/15/03	501	249,999
	05/16/03	556	277,444
	05/17/03	623	310,877
	05/18/03	456	227,544
	05/19/03	563	280,937
	05/20/03	367	183,133
	05/21/03	356	177,644

FIGURE 8.48 This is an example of a page that lacks useful row headers.

Get the picture? Repeating the current row header at the beginning of each new page costs you nothing but benefits your readers a great deal.

Summary at a Glance

Topic	Practices
Delineating Columns and Rows	• Use white space alone whenever space allows. • When you can't use white space, use subtle fill colors. • When you can't use fill color, use subtle rules. • Avoid grids altogether.
Arranging Data	• Columns or rows • Arrange a set of categorical subdivisions across separate columns if they are few in number, and the maximum number of characters in those subdivisions is not too large. • Arrange times-series subdivisions horizontally across separate columns. • Arrange ranked subdivisions vertically down the rows. • Groups and breaks • Use just enough vertical white space between groups to make breaks noticeable. • Repeat column headers at the beginning of each new group. • Keep table structure consistent from group to group. • When groups should be examined independently, start each on a new page. • Column sequence • Place sets of categorical subdivisions that are arranged down the rows of a single column to the left of the quantitative values associated with them.

Topic	Practices
Arranging Data *(continued)*	• Place sets of categorical subdivisions that have a hierarchical relationship from left to right to reflect that hierarchy. • Place quantitative values that were calculated from another set of quantitative values just to the right of the column from which they were derived. • Place columns containing data that should be compared close together. • Value sequence • Whenever categorical subdivisions have a meaningful order, sort them in that order.

Formatting Text	• Orientation
	   • Avoid text orientations other than horizontal, left to right.
	
	 • Alignment
	   • Align numbers to the right, keeping the decimal points aligned as well.
	   • Align dates however you wish, but maintain a consistent number of characters or digits for each part of the date.
	   • Align all other text to the left.
	   • Center non-numeric items if they all have the same number of characters, and the number of characters in the header is significantly greater.
	
	 • Number formatting
	   • Place a comma to the left of every three whole-number digits.
	   • Truncate the display of whole numbers by sets of three digits whenever numeric precision can be reduced to the nearest thousand, million, billion, etc.
	   • When negative numbers are enclosed in parentheses, keep the negative numbers themselves right aligned with the positive numbers.
	
	 • Date Formatting
	   • Express months either as a two-digit number or a three-character word.
	   • Express days as two digits.
	   • Express years consistently as either two or four digits.
	
	 • Number and date precision
	   • Do not exceed the required level of precision.
	
	 • Font
	   • Select a font that is legible, and use the same font throughout the table.
	
	 • Emphasis and color
	   • Boldface, italicize, or change the color of fonts when useful to group or highlight.

Summarizing Values	• Make columns containing group summaries visually distinct from detail columns. • Consider placing summaries in the group header if the information extends down multiple pages.
Giving Page Information	• Repeat column headers at the top of each page. • Repeat current row headers at the top of each page.

PRACTICE IN TABLE DESIGN

Nothing helps learning take root like practice. You will strengthen your developing expertise in table design by working through a few real-world scenarios.

Exercise #1

The following table has been prepared for a regional sales manager for the purpose of tracking the quarter-to-date performance of her sales representatives, including their relative performance. Given this purpose, take a look at the table, and follow the instructions below.

Quarter-to-Date Sales Rep Performance Summary
Quarter 2, 2011 as of March 15, 2011

Sales Rep	Quota	Variance to Quota	% of Quota	Forecast	Actual Bookings
Albright, Gary	200,000	-16,062	92	205,000	183,938
Brown, Sheryll	150,000	84,983	157	260,000	234,983
Cartwright, Bonnie	100,000	-56,125	44	50,000	43,875
Caruthers, Michael	300,000	-25,125	92	324,000	274,875
Garibaldi, John	250,000	143,774	158	410,000	393,774
Girard, Jean	75,000	-48,117	36	50,000	26,883
Jone, Suzanne	140,000	-5,204	96	149,000	134,796
Larson, Terri	350,000	238,388	168	600,000	588,388
LeShan, George	200,000	-75,126	62	132,000	124,874
Levensen, Bernard	175,000	-9,267	95	193,000	165,733
Mulligan, Robert	225,000	34,383	115	275,000	259,383
Tetracelli, Sheila	50,000	-1,263	97	50,000	48,737
Woytisek, Gillian	190,000	-3,648	98	210,000	186,352

List each of the problems that you see in the design of this table:

Suggest a solution to each of these problems:

Exercise #2

The following table is used by mortgage brokers to look up the mortgage rates offered by several lenders. Brokers use this when they need to know the current rates offered by a particular lender for all of its loan programs. Given this purpose, take a look at the table, and follow the instructions below.

Mortgage Loan Rates
Effective September 1, 2011

Loan Type	Term	Points	Lender	Rate
Adjustable	15	0	ABC Mortgage	6.0%
Adjustable	15	0	BCD Mortgage	6.0%
Adjustable	15	0	CDE Mortgage	6.0%
Fixed	15	0	ABC Mortgage	6.25%
Fixed	15	0	BCD Mortgage	6.75%
Fixed	15	0	CDE Mortgage	7.0%
Adjustable	30	.5	ABC Mortgage	6.125%
Adjustable	30	.5	BCD Mortgage	6.25%
Adjustable	30	.5	CDE Mortgage	6.5%
Fixed	30	.5	ABC Mortgage	6.5%
Fixed	30	.5	BCD Mortgage	7.0%
Fixed	30	.5	CDE Mortgage	7.25%
Adjustable	15	1	ABC Mortgage	5.675%
Adjustable	15	1	BCD Mortgage	5.675%
Adjustable	15	1	CDE Mortgage	5.75%
Fixed	30	1	ABC Mortgage	6.5%
Fixed	30	1	BCD Mortgage	6.5%
Fixed	30	1	CDE Mortgage	7.0%
Adjustable	15	1	ABC Mortgage	5.675%
Adjustable	15	1	BCD Mortgage	5.675%

List each of the problems that you see in the design of this table:

Suggest a solution to each of these problems:

Exercise #3

The following table is used by the manager of the marketing department to examine the previous year's expenses in total and by quarter for each type of expense. What you see below is the top portion of a page that is several pages into the table. The marketing manager finds this table frustrating. Can you help him out? Take a few minutes to respond to the instructions below:

2011 Marketing Department Expenses

Quarter	Transaction Date	Expense Type	Expense
	9/28/2011	Software	3837.05
	9/28/2011	Computer Hardware	10873.34
	9/29/2011	Travel	2939.95
	9/30/2011	Supplies	27.53
Qtr 4	10/1/2011	Supplies	17.37
	10/1/2011	Postage	23.83
	10/3/2011	Computer Hardware	3948.85
	10/3/2011	Software	535.98
	10/3/2011	Furniture	739.37
	10/3/2011	Travel	28.83
	10/4/2011	Entertainment	173.91
	10/15/2011	Travel	33.57
	10/16/2011	Membership Fees	395.93
	10/16/2011	Conference Registration	2195.00

List each of the problems that you see in the design of this table:

Suggest a solution to each of these problems:

Exercise #4

It's now time to shift your focus closer to home. Select three tables that are used at your place of work. Make sure that at least one of them is a table that you created. For each of the tables, respond to the following instructions:

Table #1

List each of the problems that you see in the design of this table:

Suggest a solution to each of these problems:

Table #2

List each of the problems that you see in the design of this table:

Suggest a solution to each of these problems:

Table #3

List each of the problems that you see in the design of this table:

Suggest a solution to each of these problems:

Exercise #5

This exercise asks you to design a table from scratch to achieve a specific set of communication objectives. You may construct the table using any relevant software that is available to you, such as spreadsheet software. Imagine that you are a business analyst who was asked to assess the sales performance of your company's full line of 10 products during the preceding 12 months. During the course of your analysis, you discovered that the top two products account for more than 89% of total revenue and 95% of total profit. You believe it would be beneficial to either discontinue or sell off some of the worst-performing products. To make the case, you've decided to design a table that clearly communicates your findings. The table below provides the raw data that your findings were based on, which you can expand through calculations and arrange however you choose to construct your table. Dollars have been rounded to and expressed in thousands:

Product	Units Sold	Ext Cost (000s)	Ext Revenue (000s)
A	136	$3	$7
B	119	$59	$132
C	2,938	$7	$40
D	8	$54	$92
E	4,873	$387	$402
F	25,750	$760	$1,957
G	1,837	$395	$602
H	3	$15	$20
I	13,973	$3,298	$9,266
J	93	$2	$2

Once you've completed your table, take a few minutes to describe its design in the space below, including your rationale for each design feature:

Exercise #6

You support the Vice President of Sales. She has asked for a report that she can use during presentations, which she'll be making separately to each of the six sales regions during a two-week tour of field operations. She won't display the report during these presentations. She just wants something that she can refer to when questions arise about the following areas of performance in the current quarter in each of the regions (Western U.S., Central U.S., Eastern U.S., Canada, Europe, and Asia):

- Sales revenue
- Average revenue per salesperson
- Regional percent of total revenue
- Revenue by product type (shirts, pants, dresses, skirts, and coats)
- Average order size
- Expenses

She must be prepared to answer questions about the individual performance of regions and their comparative performance as well. Design a new table for the Vice President using the following data:

Region	Number of Salespeople	Number of Orders	Expenses	Product	Revenue (U.S. $)
Western U.S.	36	4,599	11,944,850	Shirts	4,537,397
				Pants	6,738,453
				Dresses	8,503,942
				Skirts	10,376,432
				Coats	12,503,954
Central U.S.	15	2,942	3,920,940	Shirts	2,938,434
				Pants	4,682,776
				Dresses	6,039,461
				Skirts	8,239,484
				Coats	4,239,443
Eastern U.S.	24	4,112	7,135,251	Shirts	3,839,221
				Pants	5,123,044
				Dresses	7,270,982
				Skirts	9,103,845
				Coats	8,640,293
Canada	52	5,447	17,534,716	Shirts	5,988,343
				Pants	7,028,474
				Dresses	9,253,400
				Skirts	11,364,033
				Coats	17,938,444
Europe	90	7,553	40,988,486	Shirts	8,102,943
				Pants	10,384,302
				Dresses	12,982,833
				Skirts	14,135,203
				Coats	30,299,323
Asia	22	3,047	6,360,875	Shirts	2,384,332
				Pants	2,543,343
				Dresses	5,944,832
				Skirts	1,938,843
				Coats	1,323,928

You can find answers to the exercises in Appendix G, *Answers to Practice in Table Design*.

9   GENERAL GRAPH DESIGN

The visual nature of graphs requires a number of unique design practices. The volume and complexity of quantitative information that you can communicate with a single graph are astounding but only if you recognize and avoid poor design practices that would undermine your story.

Because of their visual nature, graphs tap into the incredible power of visual perception to communicate quantitative information. When the story that you wish to tell is contained in the data's patterns, trends, and exceptions; or when it depends on your audience's ability to compare entire series of values to one another (e.g., monthly domestic sales for the entire year compared to international sales), a graph will do the job best, but only if you avoid far-too-common design pitfalls.

We've already covered the aspects of quantitative communication that apply to both tables and graphs. None is more important to the design of graphs than the fundamental principle that was stated so eloquently by Edward Tufte: "Above all else show the data."[1] Quantitative stories reside in the facts, not in the containers that we use to present them. The general practice of highlighting the data and subduing all else is even more important in the design of graphs than in the design of tables. Tables are a bit more forgiving of visual design flaws because tables encode data through the use of verbal language (i.e., text), visually displayed. Graphs, in contrast, encode data as visual objects. These objects must be prominent, accurate, and clear.

1. Edward R. Tufte (2001) *The Visual Display of Quantitative Information,* Second Edition. Graphics Press, page 92.

Two fundamental principles of quantitative communication apply exclusively to graphs:

- Maintain visual correspondence to quantity.
- Avoid 3D.

Both principles are firmly rooted in practical concerns; you can wreak havoc on communication if you ignore these principles.

Maintain Visual Correspondence to Quantity

You can only use two attributes of visual perception to encode quantitative information in a way that can be easily and accurately interpreted: length and 2-D position. Quantitative values in graphs are either encoded visually as length in the form of bars or boxes or as 2-D position in the form of points and lines. Other visual attributes are either not perceived quantitatively at all (e.g., hue) or not well enough (e.g., 2-D area and color intensity) to justify their use for quantitative encoding when length and 2-D position are available.

A bar that is twice as long as another is perceived as having twice the quantitative value. Visual objects that encode quantitative values in graphs are interpreted by means of a scale line along the vertical or horizontal axis. When a bar that is twice as long as another corresponds to a value of two on the scale line, visual perception alone tells us that the value of the shorter bar is one, or very close to it. If the shorter bar actually corresponds to a value of 1.75 or 0.5, something is amiss.

People sometimes intentionally manipulate graphs to mask the truth contained in numbers. Darrell Huff, in his 1954 classic *How to Lie with Statistics*,[2] was one of the earliest to express this concern. Advertisements are notorious sources of deliberately misleading graphs, but deception is not confined to advertising. You'll be faced many times with the temptation to manipulate graphs to give your case more strength than it deserves based on the actual numbers. Given the understanding of visual design that you are developing by reading this book, you will be even better equipped to manipulate visual design to exaggerate or hide the truth. It's easy to rationalize little design manipulations here and there to shade the truth slightly for a just cause. Be aware, though, that this manipulation does not qualify as design for communication. The goal of design for communication is always to promote an accurate understanding of the truth.

2. Darrell Huff (1954) *How to Lie with Statistics*. W. W. Norton & Company.

Here's a simple illustration of the potential for deliberate misinformation:

FIGURE 9.1 These two graphs display the same information in dramatically different ways, producing radically different messages.

The graph on the left has been deliberately manipulated to make an increase in sales from $19,500,000 in July to $19,560,000 in December, which is an increase of less than one-third of 1%, look like an increase of more than 200%. The graph on the right more accurately presents the data. Do you see the specific aspects of the graph on the left that were used to exaggerate the increase in sales? Take a moment to see how many you can find, and list them in the margin to the right.

· · · · · · ·

Five design characteristics of the graph on the left give the false impression that sales have risen dramatically from July to December:

1. The scale on the Y axis does not start at zero. Rather, it starts at $19,475,000 and extends only to $19,560,000, thus making minor changes in sales appear extreme.

2. The plot area of the graph is taller than it is wide. This dramatically increases the slope of the line.

3. The line is green. The color green carries the meaning of growth and health in English-speaking cultures and dollars in the United States, so it reinforces the positive spin of the message. Also, placing a bright green line on a black background makes it pop with visual impact.

4. The highest value—the final value of $19,560,001—is set as the top of the scale. This gives the green line the appearance of extending right off the top of the graph.

5. Placing the boldfaced axis label Millions in the prominent upper left position near the title "Sales are Skyrocketing" suggests that they are increasing by millions.

This design certainly exaggerates the good news about sales, but I've seen much worse. Can you think of any additional design changes that could be made to further hide the truth?

. . . .

Here's one that I've seen:

FIGURE 9.2 This is an extreme example of intentional deceit through graph design.

Notice the changes? Values along the Y axis have been removed, and only the final data point has been labeled. Without at least one more value on the scale, there is absolutely no way to know the extent of the increase. The single value of $19,560,001, combined with the characteristics we've already discussed, together suggest a huge rate of increase. By making the graph 3D and manipulating the angle, I could exaggerate the increase even more, which is done all the time.

Now, back to the principle that prompted our journey through the dark alleys of visual obfuscation. A quantity that is visually encoded in a graph should match the actual quantity that it represents. Two specific design practices will help you honor this correspondence:

- Make the distance between tick marks on a scale line correspond to the differences in the values that they represent.
- Generally include the value zero in your quantitative scale, and alert your readers when you don't unless you're confident that they won't be misled.

Correspondence to the Tick Marks

You should always keep the distance between tick marks on a scale line consistent with the difference in the quantitative values that they represent. Software that generates graphs for you based on specified sets of values automatically enforces this practice. If the tick marks represent the values 1, 2, 3, 4, and 5, they will be positioned an equal distance from one another. If you ever produce graphs without the aid of graphing software, you should be sure to honor this practice. Approaching this from the opposite perspective, if you have a set of tick marks that are positioned at equal distances from one another, the values that you use to label them should also represent equal numeric intervals. Never place a gap in the values, such as in consecutive tick marks labeled as 1, 2, 7, 8, and 9, even if there are no values in the graph that fall within the missing range. To do so would undermine the graph's visual integrity.

You may recognize that these tick marks would not be equidistant if you were using something other than a standard scale, such as a logarithmic scale. We'll look at the special qualities and uses of logarithmic scales a little later.

Even if you indicate a break in the quantitative scale where a section of values has been eliminated, your readers could still be easily misled.

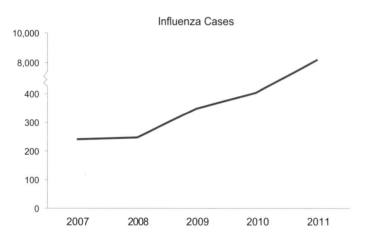

FIGURE 9.3 Scale breaks can be misleading.

Despite the fact that the scale starts at zero, the increase in influenza cases from 2010 to 2011 is underrepresented to a huge degree because of the scale break between 400 and 8,000 along the Y axis. Here's how the same values appear with a proper scale.

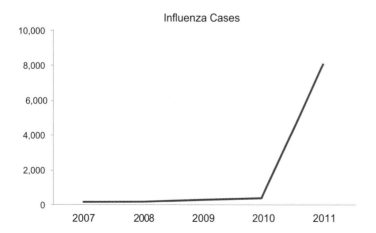

FIGURE 9.4 With a proper scale, the dramatic rise from 2010 to 2011 in influenza cases is striking.

Something is missing in this graph, however, that you might want your readers to see: the pattern of changes that occurred from 2007 through 2010. To accommodate the high number of cases in 2011, the graph's scale now forces all other values into a tiny space near the bottom, which makes the line appear almost flat during that period. How can we show the earlier pattern of change and yet still tell the more important story that the number of cases dramatically increased from 2010 to 2011? Can you think of a solution?

. . . .

To tell this entire story, two graphs are needed, such as the following.

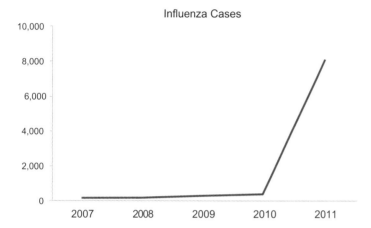

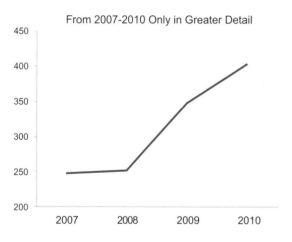

FIGURE 9.5 Two graphs are needed to tell all aspects of this story clearly.

It's important to know that quantitative stories can often only be told with more than one graph. Nothing is gained by attempting to squeeze into a single graph what can be more effectively presented in several.

Zero-Based Scales

When you set the bottom of your quantitative scale to a value greater than zero, differences in values will be exaggerated visually in the graph. Usually, you should avoid starting your graph with a value greater than zero, but when you need to provide a close look at small differences between large values, it's appropriate to do so. When you do so, alert your readers to the fact if you have any doubt that they'll notice. Perhaps you observed that the scale in the lower graph in *Figure 9.5* doesn't start at zero. Because the same information was already shown using a zero-based scale in the upper graph, the fact that the scale was adjusted in the lower graph would be hard to miss. If the sales manager of a company with the subtly rising sales that we examined in *Figures 9.1* and *9.2* wanted to examine that increase in great detail, however insignificant it might be as a percentage increase, the following graph would make this possible, but textual alerts similar to those shown in red might be needed.

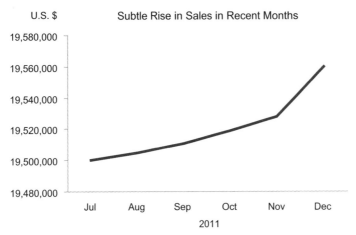

FIGURE 9.6 This is an example of an exception to the zero-based scale, illustrating how such an exception can be clearly noted to prevent misunderstanding.

Never eliminate zero from the quantitative scale when bars are used to encode the values, however. Why? Because a bar encodes quantitative value primarily through its length, and, without zero as the base, the length will not correspond to its value. In the following software ad, which I clipped from a magazine, how much greater is customer loyalty to MicroStrategy than Cognos Powerplay?

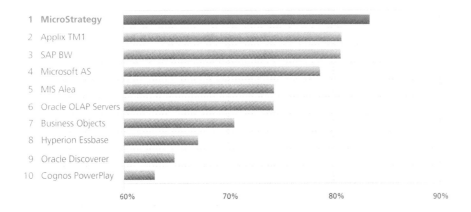

FIGURE 9.7 This graph misrepresents the values by starting the scale at 60%.

The MicroStrategy bar appears to be more than six times greater in length than the Cognos PowerPlay bar. The difference between the values, however, is about 83% versus 63%—quite a different story. Here's the same information, properly displayed:

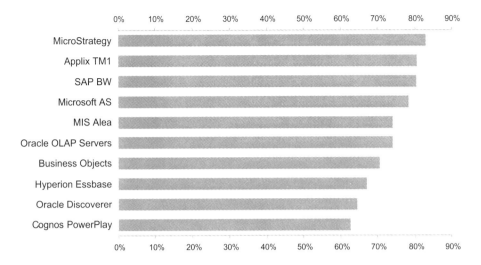

FIGURE 9.8 This graph displays the values in *Figure 9.7* properly.

When a graph represents both positive and negative numbers, zero will not mark the bottom of the scale, but it will still represent the base from which all values extend. The following two graphs contain the same set of positive and negative values. The graph on the right correctly displays zero as the base of its scale from which bars extend upwards for positive values and downwards for negative values, but the one on the left mistakenly sets the base to slightly below the lowest value, resulting in a confusing and misleading representation of the values.

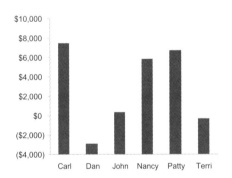 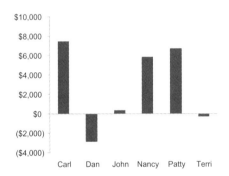

FIGURE 9.9 Both of these graphs display both positive and negative numbers. The graph on the right correctly sets zero as the base of its scale at the point where the X axis intersects the Y axis. The graph on the left incorrectly sets the base to slightly below the lowest value.

Avoid 3D

When 3D is used in graphs, it takes one of two possible forms:

- The addition of a third dimension of depth to objects (e.g., bars) that are used to encode quantitative values, without the addition of a third quantitative scale.
- The addition of a third dimension of depth to the overall graph with an associated quantitative scale (the Z axis).

Neither form is effective, but the reasons are entirely different.

Data Objects with 3-D Depth

We're using four objects to encode quantitative values in graphs: points, bars, lines, and boxes. The addition of depth to a value-encoding object does not affect the object's value. Add depth to a series of bars, and what do you have? Nothing more than bars that now occupy more space and are harder to tie to values along the scale. If you add depth to value-encoding points, like dots and squares, you get spheres and cubes that represent the same values as before, but now their depth makes it harder to align them accurately with the scale. 3-D versions of lines look like thick ribbons and suffer from the same problems.

Here are four variations of the same graph, three of which have 3-D effects added to the bars:

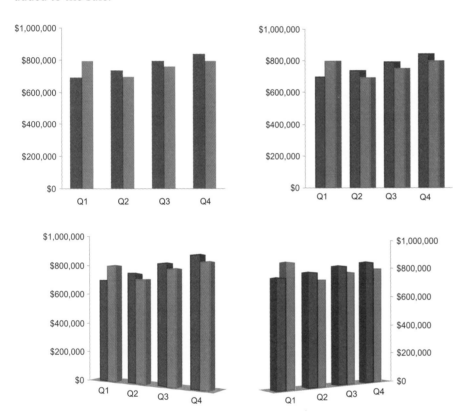

FIGURE 9.10 These four examples display the same values using bars in four different ways, three of which incorporate 3D.

Which graph is easiest to read? When shown all of these at once, the answer is obvious, isn't it? I was careful in the graphs above to keep the 3-D effects simple. If we take advantage of the many options that most software provides, we can bury truth in visual effects. In the following example, I've manipulated perspective and angles to make a steady increase in expenses from $100,000 to $121,000 look like a flat series of consistent values.

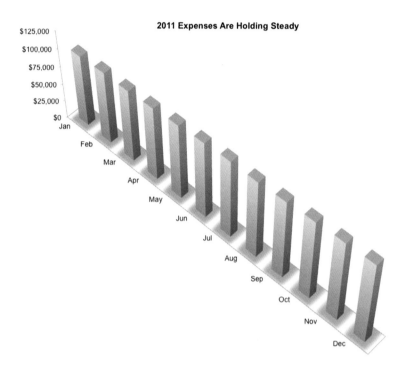

FIGURE 9.11 3-D effects are
sometimes used to tell lies.

Most software makes it far too easy and tempting to add a third dimension to
objects in graphs. This functionality is thrown in because people expect it, not
because it's useful. It is far better to impress your readers with graphs they can
easily understand and use than graphs that look like video games and are
difficult to interpret.

Remember the data-ink ratio. The addition of 3D to value-encoding objects
adds ink but not data. That is, it adds meaningless visual content that your
readers must take in and process, resulting in nothing but wasted time and
effort.

Graphs with 3-D Depth

A third dimension of depth may be added to an entire graph through the use of
a third axis, conventionally called the Z axis. The Z axis may be used for either a
categorical or a quantitative scale. A categorical scale along the Z axis allows you
to add another set of categorical items that extend back along the axis, accompa-
nied by related rows of quantitative values. A quantitative scale along the third
axis can display a third quantitative variable in a scatter plot. In theory, this is a
valid way to include more information in a graph. In practice, with rare exceptions, it is simply too hard to read. Simulating 3-D space on a 2-D surface works
nicely for paintings or technical illustrations but almost never for graphs.

A few examples will vividly illustrate this point. Let's start with the same data
that we examined earlier as the dark gray bars in *Figure 9.10.*

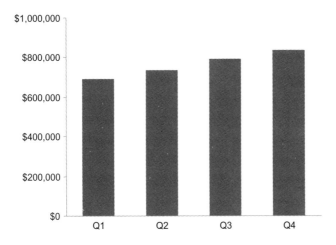

FIGURE 9.12 This is an example of a simple 2-D graph.

So far we have a very simple 2-D graph. Now let's say that we want to display these quarterly bookings by the four sales regions of North, East, South, and West. To do so, we could encode each region as a different hue and keep the graph 2D, as follows:

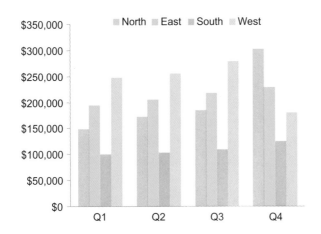

FIGURE 9.13 This 2-D graph has two sets of categorical items: quarters along the X axis and sales regions encoded as different hues.

This is still fairly easy to read. Rather than using hue to encode the four sales channels, we could instead add a Z axis to the graph, making it 3D, and display the sales channels along that axis.

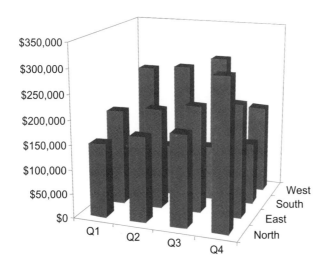

FIGURE 9.14 This is a 3-D graph, with sales in dollars along one axis, quarters along another, and sales regions along a third.

This is a very simple example of a 3-D graph with two categorical scales (quarters and regions) on one quantitative scale (dollars). What do you think? Does it work? Examine it for a moment, attempting to read and compare its values. Try to describe what makes this graph difficult to read.

. . . .

When a third dimension is added to a graph, adjustments are usually made to the graph automatically by software—tilting, rotating, and adding perspective—to make its data more visible. A 3-D display like this is called an *axonometric projection*. The previous example was tilted down 15 degrees, rotated clockwise 20 degrees, and given 30 degrees of perspective. These variables can be altered in an effort to make the graph easier to read. Even though the graph has been tilted and rotated in an attempt to make the rows of bars more visible, some bars will always remain partially or entirely hidden. Also, it's nearly impossible to line the bars up with values along the quantitative scale.

Software that generates 3-D graphs often includes grid lines on the walls in an effort to make the quantitative values easier to align with the scales lines. Here's the same graph as before with the addition of these features, along with black borders around the bars to more clearly delineate them.

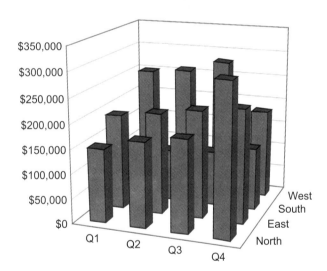

FIGURE 9.15 This is a 3-D graph that has been enhanced in an effort to make the values easier to read through the use of grid lines on the walls and borders around the bars.

Even though this is a fairly simple graph, these enhancements still don't solve the problems. Software vendors sometimes argue that this problem can be solved by rotating the graph to see bars that are hidden. The problem with this approach is that any new perspective will reveal some bars and hide others, never allowing us to see and compare all the values at once, which is a key benefit of graphs. Changing from the use of bars to lines to encode the data doesn't fix the problem either, as you can see in the following example:

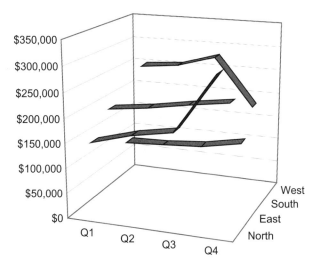

FIGURE 9.16 This graph displays the same data as *Figure 9.15* but this time using lines to encode the values.

Which of the lines in this graph represents the south region? When I ask this question in classes, fewer than 50% of my students answer correctly. The lowest of the four lines represents the south region, but this isn't at all obvious, is it? Support components called *drop lines* were invented to help us locate data objects in relation to scales along axes, especially in 3-D graphs, but they clutter the graph and reduce its interpretation to a slow series of look-ups.

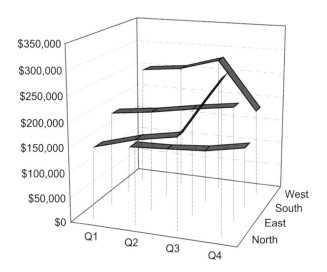

FIGURE 9.17 This graph is the same as the one in *Figure 9.16* with the addition of drop lines.

So far we've only examined the association of a categorical scale with the third axis. The problems don't get any better when the third axis is used for a quantitative scale. Imagine a scatter plot that correlates employee salaries in dollars along one axis, tenure on the job in years along another axis, and level of education in years along the final axis. It's too difficult to tell where the points are positioned along the third axis.

3-D renderings of quantitative information rarely work. Don't sacrifice effective communication through the use of 3-D fluff. Even when you are driven by a sincere desire to give your readers more information by using a third dimension, there are better ways to realize these good intentions. One effective technique is to use multiple related graphs in a series, which we'll explore in Chapter 11, *Displaying Many Variables at Once*.

Summary at a Glance

- Encode quantities to correspond accurately to the visual scale.
  - Keep the distance between tick marks on a scale line consistent with the difference in the quantitative values that they represent.
  - In most cases include the value zero in your quantitative scale, and alert your readers when you don't. Always start the quantitative scale at zero when you use bars to encode the values.
- Avoid 3-D displays of quantitative data.

10 COMPONENT-LEVEL GRAPH DESIGN

Several visual and textual components work together in graphs to present quantitative information. If these components are out of balance or misused, the story suffers. For each component to serve its purpose, you must understand its role and the design practices that enable it to fulfill its role effectively.

Graphs are constructed from components. Most but not all components represent data. Data components can be divided into two groups: primary components (points, bars, lines, and boxes) and components that serve secondary roles (scales, trend lines, tick marks, and so on). This chapter is organized into the following sections:

- Primary data component design
- Secondary data component design
- Non-data component design

Before jumping in, let's get our terminology straight. Here's a diagram that shows most of the terms used to describe graph components in this chapter:

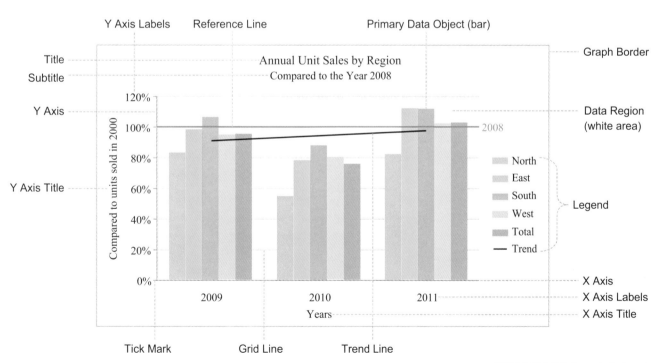

FIGURE 10.1 This figure is designed only to label the parts of a graph, not to illustrate best practices in graph design.

Primary Data Component Design

Graphs that are properly designed can be decoded as an act of visual perception, tapping into our powerful ability to detect patterns in quantitative information that is displayed visually. Quantitative values are encoded in graphs as points, bars, lines, or boxes. To assign precise values to these objects, most graphs use quantitative scales along one or more axes, labeled with numbers and tick marks. Categorical items are primarily encoded as visual attributes of the points, bars, lines, or boxes (e.g., color or location in 2-D space). Text labels are used to assign categorical items to these attributes, either along the axes or by means of legends. Just as in tables, text may also be used in graphs to introduce the display and provide explanations, reinforce and highlight particular items, sequence elements, recommend responses in the form of decisions and actions, and pose important questions.

Because we've already spent time in Chapter 6, *Fundamental Variations of Graphs* examining how points, bars, lines, and boxes encode values, we'll move on now to 1) the finer details of their design, 2) ways they can be combined for specific purposes, and 3) a few data components that we haven't examined so far, such as tick marks and reference lines.

Points

When you encode quantitative values as points (dots, squares, triangles, and so on), all design practices address a single primary objective: keep points easy to perceive. Let's walk through a few examples of graphs that illustrate various perceptual problems that can plague points.

This first example uses different point shapes to distinguish two sets of values. However, it's harder than it should be to see the two groups as distinct. Take a moment to describe, in the margin to the right, the problem as you see it:

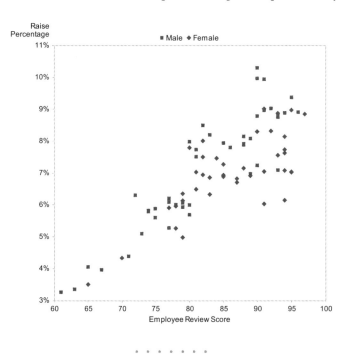

FIGURE 10.2 This is an example of points that suffer from a particular perceptual problem.

The problem here is that the points are so small that their distinct orientations—vertically oriented squares for the men versus squares rotated 45° (i.e., diamonds) for the women—cannot be distinguished easily. We can remedy this lack of perceptual distinction between the two sets in a number of ways. Take a minute and try to identify one or two ways to fix this problem.

One simple solution involves enlarging the points. Nothing is different about the following example except that the points are bigger:

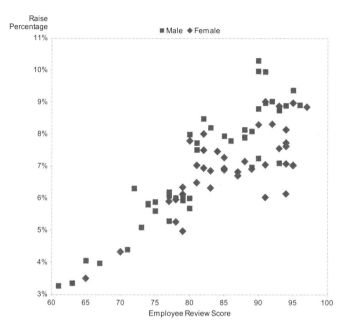

FIGURE 10.3 This example demonstrates that sets of points can be made more distinct by enlarging them.

Enlarging the points certainly helps, but another minor change will help even more: remove their fill color. Differences between simple shapes, or in this case between a single shape that has been oriented in two ways (the diamonds are merely squares rotated 45 degrees), can be more easily distinguished when the fill color has been removed, leaving only their borders, as you can see in this next example:

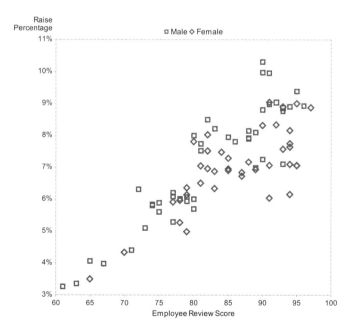

FIGURE 10.4 This example demonstrates that sets of points can be made more distinct by removing fill color.

If you want a method that doesn't involve enlarging the points, select shapes that are more visually distinct. An example appears on the next page.

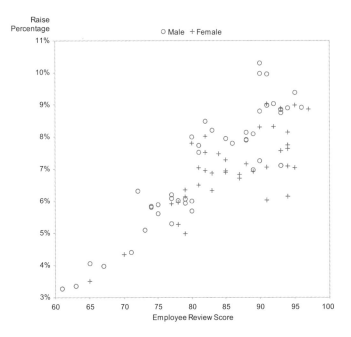

FIGURE 10.5 This example demonstrates that sets of points can be made more distinct from one another by using shapes that are very different.

Some shapes that are easiest to distinguish from one another are the following:

○ □ △ + ✕

So far we've used different point shapes to distinguish groups, but there is a preattentive visual attribute that we can distinguish better than shapes. What do you think it is?

· · · · · · ·

The answer is hue. In the next example, the two groups of points can be easily distinguished even though they're fairly small.

All of these shapes are now available in Excel. These symbols are easier to distinguish if you keep those with interiors (circle, square, and triangle) empty of fill color.

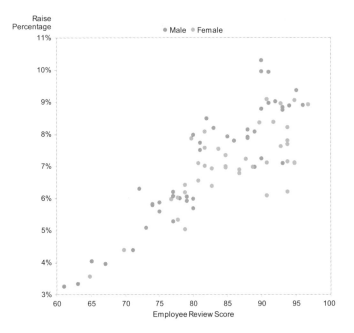

FIGURE 10.6 This example demonstrates that sets of points can be made more distinct from one another by using different hues.

Hues work best for distinguishing groups of points. Stick with hue for this purpose unless you can't because you've already used hue differences for another purpose in the graph, or you're concerned that the graph will be printed in black and white, resulting in the loss of hue distinctions.

Let's move on to another perceptual problem that can plague points, especially in scatter plots that show a large number of points. Take a look at the following example and describe in the right margin the problem that you detect:

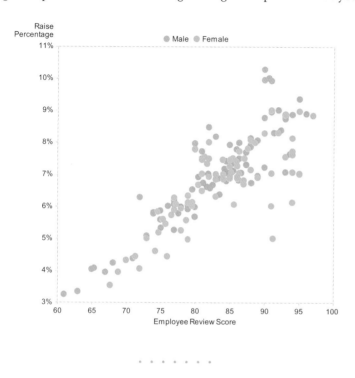

FIGURE 10.7 This example shows points that suffer from a visibility problem.

. . . .

This scatter plot contains many more points than the previous examples. The problem that it illustrates is that points can overlap, potentially causing some to be obscured completely. This problem is called *over-plotting*. When points are hidden or overlapped by other points to the degree that they can't be distinguished, incomplete and inaccurate communication results.

When points overlap somewhat, but none are entirely hidden, this can sometimes be remedied by enlarging the graph, decreasing the size of the points, or a combination of the two. These steps will reduce the overlap. If the problem persists, another technique can be used that usually fixes the problem. Knowing what you now know about visual perception, look again at the last example and see whether you can come up with any other way to remedy the problems associated with the overlapping points.

. . . .

Any luck? Your ability to recognize opportunities to improve visual perception by manipulating visual attributes should be sharpening, so you may have detected that the solid nature of the points, the fact that they are filled with color, reduces their discrete visibility where overlapping occurs. Let's see what happens when we remove the fill colors, leaving only the outlines of the dots:

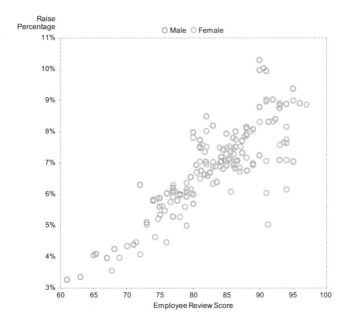

FIGURE 10.8 This example demonstrates that the problem of overlapping data points can be reduced by removing the fill color from the points.

This makes a big difference. When the points have transparent interiors, we can more readily see when points overlap. This same technique works with a variety of point shapes, not just dots.

Bars

Several characteristics of bars deserve attention. We'll consider each of the following:

- Orientation
- Proximity
- Fills
- Borders
- Base value

ORIENTATION

In this context, orientation refers to the direction that the bars run, either horizontally from left to right or vertically from bottom to top. Each orientation has advantages in particular circumstances.

Horizontal bars are the best choice when it would be difficult to get categorical labels to fit side by side under vertical bars. In the following example, even though the names of these sales representative names are not abnormally long, placing them side by side to label vertical bars results in an awkwardly wide graph:

FIGURE 10.9 This graph illustrates the problem with vertical bars when their categorical labels are long.

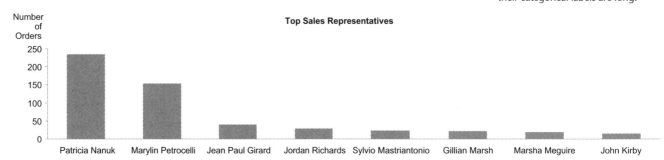

Perhaps you could make this work, but what if the graph displayed twenty sales representatives rather than eight?

You might be thinking that this problem could be solved by orienting the names at an angle or vertically, as in the following example.

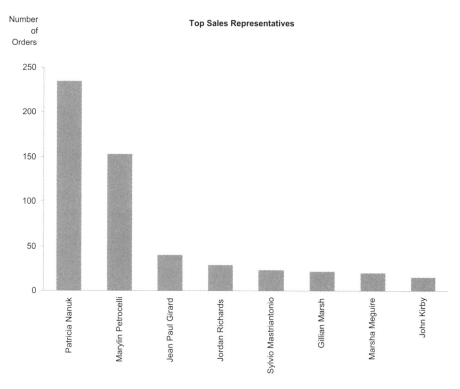

FIGURE 10.10 This graph illustrates the problem that results when we associate long categorical labels with vertical bars by orienting them vertically.

This solves the horizontal space problem but makes the names hard to read. When you must use vertical bars and the categorical labels are too long to fit side by side, opt for an upward-sloping angle of 45° or less rather than orienting the labels vertically. This is much easier to read, as in the following example:

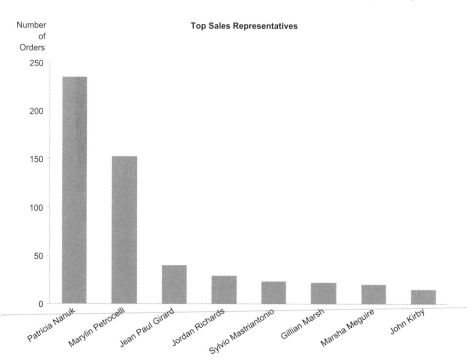

FIGURE 10.11 This graph orients the categorical labels at an angle, rather than vertically, to make them more legible. These are oriented at a 33° angle.

It is almost always better, however, to solve the problem with horizontal rather than vertical bars if you can. Notice how well this solution works.

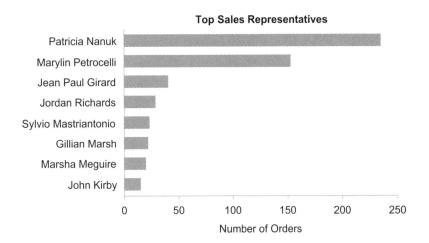

FIGURE 10.12 Horizontal bars handle long categorical labels nicely.

Long labels aren't a problem at all when you use horizontal bars. Even when labels aren't especially long, if you need to show a large number of bars in a graph, orienting them horizontally usually works best. The one exception is when you're displaying change through time (years, months, days, etc.) in a bar graph; you should almost always place time on the X axis so that it runs horizontally from left to right, which will result in vertical bars.

PROXIMITY

Now let's consider how closely bars should be placed to one another. You should always maintain a balance between the width of the bars themselves and the white space that separates them. In other words, rather than thinking of the space between the bars in absolute terms, think of it in terms of a ratio of the width of the bars to the width of the white space between them. Here are several examples of graphs with different ratios of bar width to white space. Take a look at the next five figures and determine which seems to work best:

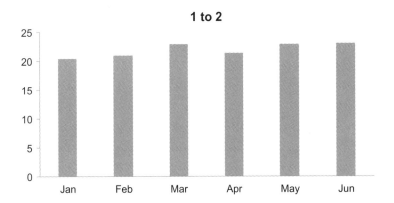

FIGURE 10.13 This is an example of a 1-to-2 ratio of bar width to intervening white space.

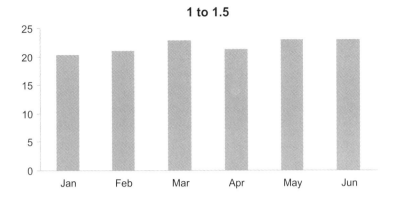

1 to 1.5

FIGURE 10.14 This is an example of a 1-to-1.5 ratio of bar width to intervening white space.

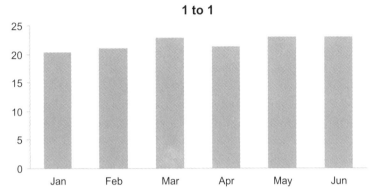

1 to 1

FIGURE 10.15 This is an example of a 1-to-1 ratio of bar width to intervening white space.

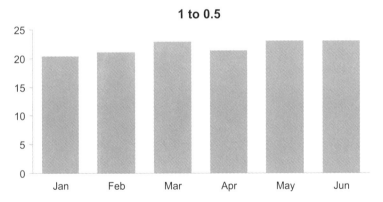

1 to 0.5

FIGURE 10.16 This is an example of a 1-to-0.5 ratio of bar width to intervening white space.

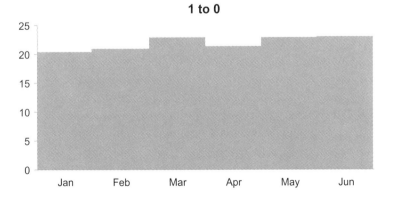

1 to 0

FIGURE 10.17 This is an example of a 1-to-0 ratio of bar width to intervening white space.

I'm not aware of any research that suggests which of these ratios works best. Personally, I prefer ordinarily to stick within the range extending from a ratio of

1:1.5 to 1:0.5 and lean toward a ratio of 1:1 as ideal. Larger ratios produce too much white space. At the other extreme, as in the 1:0 ratio example, the bars cease to be discrete, suggesting a continuous range of values that is only appropriate along an interval scale or when displaying a part-to-whole relationship. At this extreme, the unique ability of bars to display individual values as discrete is almost entirely lost.

The primary occasion when space between bars isn't needed is when several bars correspond to each categorical item labeled along the axis. In the examples below, four bars, one per region, appear side by side in each quarter.

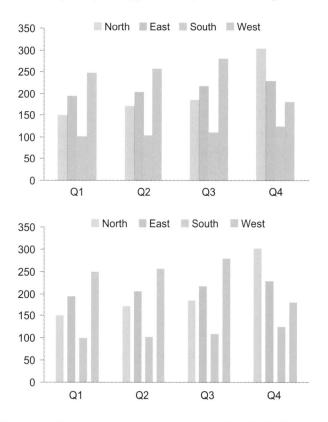

FIGURE 10.18 This graph illustrates when it is appropriate to place bars side by side without white space in between.

FIGURE 10.19 This graph shows that it's unnecessary to insert white space between bars that are associated with the same categorical item along the axis.

Allowing the bars within each quarter to touch, as they do in the upper example, reinforces the fact that they belong together as part of a group.

What about overlapping the bars? You could go further than removing white space between bars and actually overlap them. Examine the next example to see how this looks:

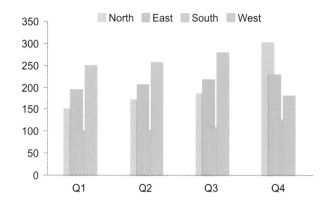

FIGURE 10.20 This is an example of overlapping bars, which illustrates the visual problems that overlapping creates.

When bars overlap in this manner, they look a bit like a jigsaw puzzle. The pink bars appear more dominant than the others because they occupy the front position. Some bars, like the orange ones, take on odd shapes. Overall, the graph is visually confusing, and thus hard to read, showing clearly why it's best to avoid overlapping bars.

FILLS

The use of fill colors or patterns in bars should follow the general design practices that we've already examined:

- Avoid the use of fill patterns (e.g., horizontal, vertical, or diagonal lines) because they create disorienting visual effects.
- Use fill colors that are clearly distinct.
- Use fill colors that are fairly balanced in intensity for data sets that are equal in importance.
- Use fill colors that are more intense than others when you wish to highlight particular values above the others.

This last practice in this list is a useful way to direct your readers' attention to particular data without being overbearing, such as by giving the south region in the following example a darker, more saturated color.

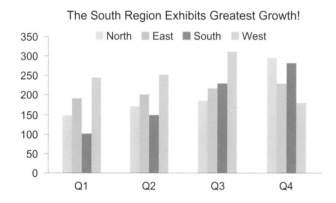

FIGURE 10.21 This is an example of using an intense fill color, in this case dark green, to highlight a particular set of values.

We should add one more practice to this list: use only one fill color per set of related values. This practice hasn't been followed in the graph on the left below:

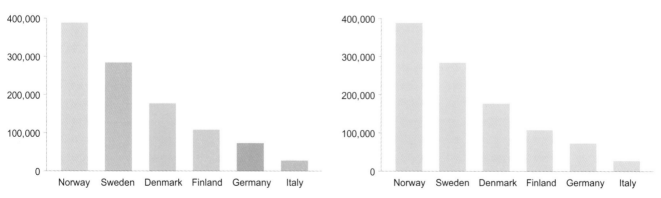

FIGURE 10.22 For a single series of values, the bars on the left differ in color and those on the right are the same.

Which one of these graphs does a better job of encouraging your eyes to compare the values? The one of the right, correct? The sameness of the color causes you to see the bars as a group, which in fact they are, and your eyes are therefore encouraged to compare them to one another. The one on the left does the opposite. Differences in color discourage us somewhat from comparing the bars. Given the fact that each bar is identified with labels along the X axis, what do the colors in the graph on the left mean? Nothing. A good general principle in graph design is to never vary anything—color, size, position, angle, etc.—unless the differences mean something. Meaningless variation unnecessarily complicates the display and sends your readers on a search for meanings that don't exist.

BORDERS

A border around a bar is only visible if the border's color is different from the fill color of the bar. The use of bar borders usually adds a visual component to the graph without adding information. Here are examples of some of the possible variations:

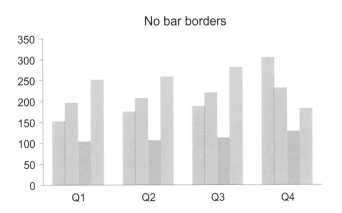

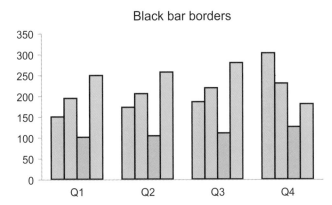

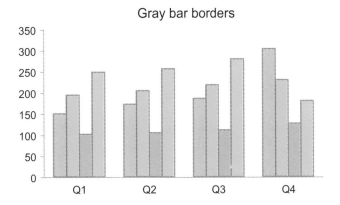

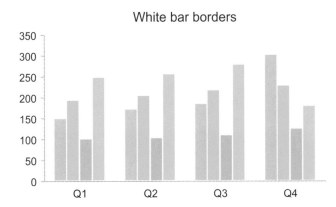

Other than for highlighting, borders around bars are only useful when bar fill colors do not sufficiently stand out against the background (e.g., light yellow bars against a white background). If you must use light-colored bars for some reason, the use of subtle borders (e.g., gray rather than black) creates sufficient separation between the bar and the background to make the bars stand out.

Just like fill colors, borders may be intentionally introduced to make particular values or sets of values stand out from the others. In the following example, attention is clearly drawn to a particular bar through the use of a border that is absent from the other bars:

FIGURE 10.23 These examples show variations in the use of borders around bars.

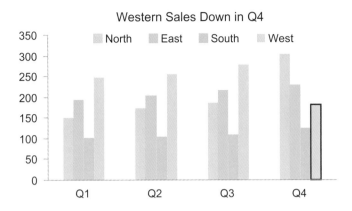

FIGURE 10.24 This example uses a border around a bar to highlight a particular value

BASE VALUE

A bar consists of two ends: the one that marks the value, called the *endpoint*, and the one that forms its beginning, called the *base*. The purpose of this section is to consider where bars should begin along a quantitative scale. Bars should begin at zero and extend from there. This is not the same as saying that bars should always begin at the bottom or left edge of the graph. The axis that the bars rest on does not intersect the other axis at its lowest quantitative value when both positive and negative values are included. In the following example, the X axis intersects the Y axis near the middle:

FIGURE 10.25 This is an example of a graph with an X axis that is not positioned at the end of the Y axis.

Whether bars extend upward or downward, to the right or to the left, they should start at zero. Otherwise, their lengths will not correspond to the values they encode.

Lines

In contrast to bars, which emphasize individual values, lines emphasize continuity and flow from one value to the next. Consequently, lines are particularly good at displaying values that change through time, featuring the overall shape of that change. With rare exceptions, lines should only be used to connect values along an interval scale (i.e., to display a time series or a distribution). In this section we'll examine ways to make lines easy to distinguish from one another and circumstances when points should be included on lines to mark the location of values.

DISTINGUISHING LINES

When a graph contains multiple sets of values, each encoded as a line, you must take care to make them visually distinct. Lines that look too much alike are hard to trace as they cross one another, as in the following example:

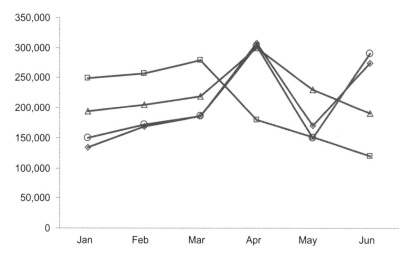

FIGURE 10.26 This graph shows lines that are not clearly distinct from one another.

In this example, the only visual differences that distinguish the lines are the distinct shapes of the points that mark the values. This distinction isn't sufficient, however, especially where lines intersect. Shapes that mark individual data points don't solve this problem because they only appear intermittently along the length of the line. Unless you want to simultaneously display the individual values as well as the overall shape of the data, you wouldn't bother to mark the individual values with points anyway, so you definitely shouldn't get into the habit of distinguishing lines by symbol shapes alone. Given what you know about the attributes of visual perception, how could you make the lines sufficiently distinct?

· · · · · · ·

The most effective method uses differences in hue or color intensity. Here are two new versions of the same graph without distinct point shapes but with distinct colors instead:

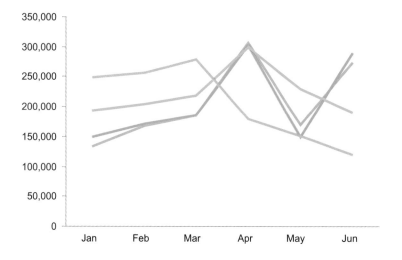

FIGURE 10.27 This graph shows hues used to make lines clearly distinct.

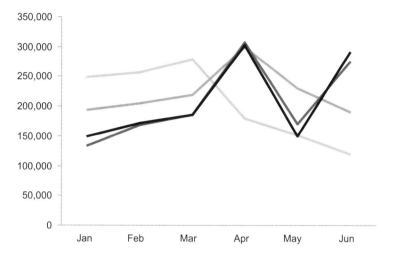

FIGURE 10.28 This graph shows different color intensities used to make lines distinct.

Hues work more effectively than color intensities, but if the graph isn't printed in color, differing shades of gray are more likely to remain distinct. One disadvantage of this approach is that the darker lines stand out more than those that are lighter, which suggests that the darker lines are more important that others. Sometimes this is an asset because it provides a way to display a ranking relationship. In the case above I've used the order of dark to light lines to rank the lines according to their values in the final month of June.

Another option that's available for distinguishing lines, which is also useful when hues can't be used, is variations in line style, as in the following example:

Although lines of significantly different shades of gray will usually look different when printed on a black and white printer, they will not necessarily look different or even be visible when copied on a black and white photocopier. When a gray line is scanned by a black and white photocopier, there is a tendency for light gray lines to become white and dark gray lines to become black. If you know your graph will be copied on a black and white photocopier, you might have to use black lines with one of the inferior encoding methods, such as differing symbol shapes or line styles.

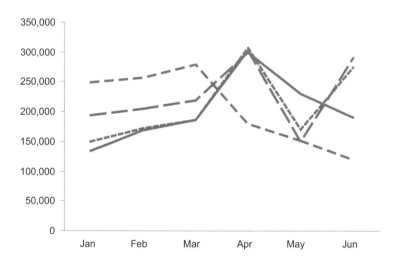

FIGURE 10.29 This graph shows line styles used in an attempt to make lines clearly distinct.

One significant drawback of this approach, as you can see, is that dashed styles break up the flow of the line, producing a jagged appearance that is not inviting to the eye. For this reason it is best to avoid varying line styles unless you can't use hue or color intensity.

POINTS OR NO POINTS

You have perhaps noticed that I don't always include data points to mark values along the line. The stories the lines tell in graphs are usually contained in the patterns that the lines form, not in precise values, so points aren't necessary. Nevertheless, there are times when it is useful to include points. The primary case is when the graph contains multiple lines that people will use to compare values at a particular point in time. For example, in the next figure, it would be difficult to know precisely where along the three lines you should pinpoint to compare values for the month of June.

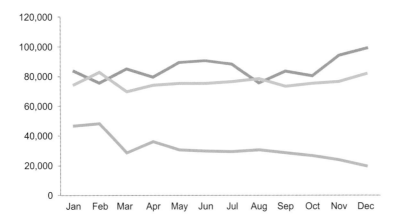

FIGURE 10.30 Without points, it is sometimes difficult to compare values on different lines at the same point in time.

With the addition of points, this task becomes easier because the points tell us precisely where to make the comparison, as you can see in the example below:

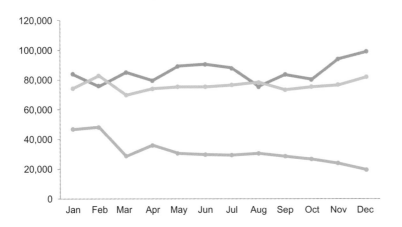

FIGURE 10.31 The addition of points makes it easier to compare values on different lines at the same point in time.

Notice that it isn't necessary to make the points large to do the job. In fact, it works best to keep them small to avoid clutter. Another way to solve this problem is to include light vertical grid lines in the graph to mark the position of each month, rather than points, as illustrated below:

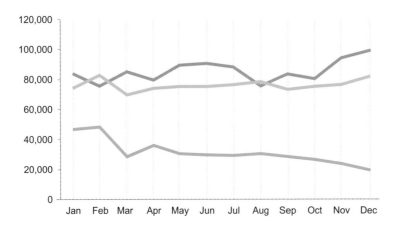

FIGURE 10.32 The addition of light vertical grid lines makes it easier to compare values on different lines at the same point in time.

Boxes

Most of the design practices that apply to bars apply to boxes as well because of the similarities between them.

- Boxes may be oriented either vertically or horizontally.
- Space between boxes that exceeds the width of a box is wasteful, making comparisons more difficult because of excessive distance between them, except when they're used in time series displays where the extra space between them is useful for showing the pattern of change through time. (I'll illustrate this in a moment.)
- When you have multiple sets of boxes in a graph, differing their fill colors works best to distinguish them.
- Borders around boxes aren't necessary unless the box's fill color does not stand out sufficiently against the background.

The one bar design practice that doesn't apply to boxes concerns the base. Bars must begin at a common base of zero on the quantitative scale, but boxes are different because they don't each display a single value as bars do. Instead, boxes show the distribution of a full set of values from low to high.

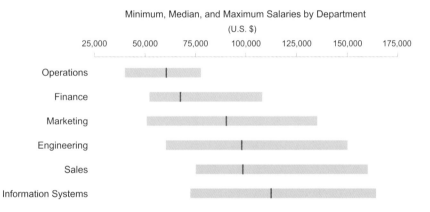

FIGURE 10.33 Unlike bars, boxes don't require that the quantitative scale begin at zero.

In this example, because the categorical labels are fairly long, I chose to use horizontal boxes, which allowed me to keep the boxes close together for easy comparison.

The two examples below both show the same series of distributions over time. Although the version on the left makes it slightly easier to compare individual distributions to one another by placing the boxes in close proximity, the increased width of the graph on the right allows it to do a better job of showing the pattern of how the mean selling price changed through time.

FIGURE 10.34 These box plots are designed to show how distributions change through time.

These two graphs were created in Excel using a line graph with high-low lines at each value, which is described in Appendix E, *Constructing Box Plots in Excel*.

As with most time-series displays, it usually works best to make the graph wider than it is tall to feature the pattern of change without exaggeration. Except when you are displaying distributions over time, keep boxes fairly close to one another so they can be easily compared.

Useful Combinations of Points, Bars, Lines, and Boxes

Points, bars, lines, and boxes are ordinarily used independently in graphs. They shouldn't be combined arbitrarily, but there are occasions when they collaborate in meaningful and effective ways. In the following example, the graph on the top combines bars and a line arbitrarily, which results in a solution that works less well than the one underneath:

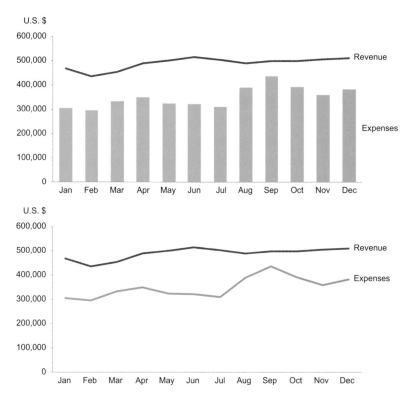

FIGURE 10.35 The top graph combines bars and a line arbitrarily whereas the one on the bottom uses two lines, which reveals patterns and supports comparisons better.

When displaying time-series data, we would only use bars if we wanted to feature values individually rather than as an entire series. For what possible reason would we want to feature the pattern of changing revenues by using a line but feature monthly expenses individually by using bars as shown in the upper graph above? Doing so not only makes it harder to see the pattern of changing revenues but also harder to see how the pattern of change in revenue compares to that of expenses. Choices between points, bars, lines, and boxes should never be arbitrary. We should always choose the means of display that can tell the story best. There are three occasions in particular when it is useful to combine different value-encoding objects:

- Combining boxes and lines to display distributions through time
- Combining bars and lines for ranked contributions to the whole
- Combining bars and points for uncluttered comparisons

BOXES AND LINES FOR DISTRIBUTIONS THROUGH TIME

The first of these combinations we've already seen. The final box plot example, *Figure 10.34*, used boxes to show individual distributions and a line to show how their center values (which in this case are means) changed through time. This combination was not arbitrary; the boxes and lines took advantage of the strength of each form of display—boxes for distributions and a line for the pattern of change through time—to tell a story that featured both.

BARS AND LINES FOR RANKED CONTRIBUTIONS TO THE WHOLE

While studying the distribution of wealth in Italy, a social scientist named Vilfredo Pareto discovered that approximately 20% of the population owned 80% of the wealth. From there, he further observed that these proportions describe many other aspects of society as well, which led him to propose the 80-20 rule. A special kind of graph that can be used to show how large amounts of something are associated with small portions of a population, called the *Pareto chart*, was named in his honor. Pareto charts combine ranking and part-to-whole relationships to tell the story of how the biggest parts of something combine to dominate the whole. In the following example, different channels through which people discover an organization are displayed in a Pareto chart.

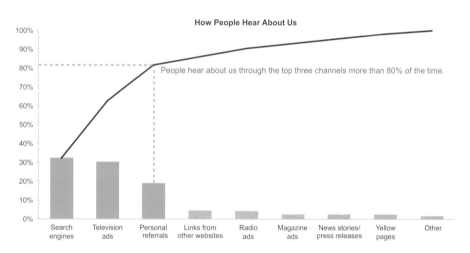

FIGURE 10.36 This Pareto chart presents, in ranked order, the channels through which people discover an organization.

A Pareto chart combines bars and a line in a useful way. The bars display a ranking relationship between parts of the whole from greatest to least, in this case channels through which people discover the organization. Bars make it possible to easily compare the contribution of individual channels. A line accumulates the same series of values to show the percentage of the whole contributed by subsets of values from high to low. A line makes sense in this case because there is an intimate connection from one value to the next as they combine to form the whole. This particular example was designed to tell the story that most people discover the organization through the three top channels. The clear message is that that the organization could increase its visibility by doing what it can to improve these three channels.

Pareto charts can be easily created in Excel simply by associating a column chart (vertical bars) with the series of individual values and a line chart with the series of cumulative values.

BARS AND POINTS FOR UNCLUTTERED COMPARISONS

Another combination that is often handy uses bars to feature a primary set of values and points to display one or more sets of values that you want to compare with the primary set. A good example involves a comparison of actual expenses to budgeted expenses, by department:

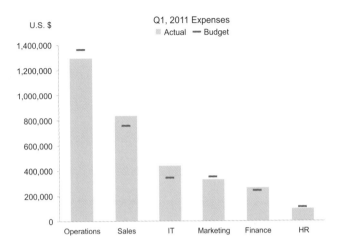

FIGURE 10.37 This graph uses bars to display a primary set of values and points to display comparative values.

In this case, actual expenses are featured, and the budget amounts are there merely to provide context in the form of a comparison that allows the viewer to assess performance. Rather than using bars for both sets of values, using bars for actual expenses only and points for budget values causes actual expenses to stand out clearly as the more important of the two. Another benefit of this design is that it is less cluttered and requires less space than a graph with two sets of side-by-side bars. Graphs of this type can be used for any comparison that features a single set of values but also provides other values for the purpose of comparison.

Graphs of this type can be created in Excel by using a column chart (vertical bars) for the primary set of values and a line chart for the other set. Display points along the line and then turn the line off, so only the points remain visible.

Secondary Data Component Design

In addition to the primary data objects that are used to encode values in graphs (points, bars, lines, and boxes), several other components of graphs also provide information. In this section, we'll look at ways to design each of the following components for maximum effect:

- Trend lines
- Reference lines
- Annotations
- Scales
- Tick marks
- Grid lines
- Legends

Trend Lines

A trend line describes the overall nature of a set of values, usually in a graph that displays correlations or time series. As such, trend lines do not represent

actual values but instead summarize their essence. In this section we'll examine
the role of trend lines that display time series and correlations.

TREND LINES IN TIME SERIES

The following time-series display includes a trend line to show the overall trend
of change during a particular 12-month period:

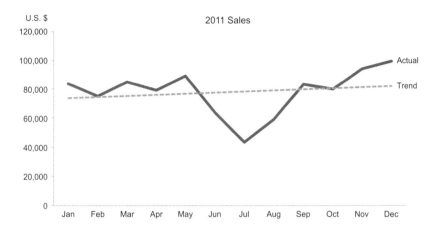

FIGURE 10.38 This graph shows a
trend line that represents the overall
pattern of sales during the year.

A trend line of this type is also called straight line of best fit or a linear trend
line. It is an attempt to draw a straight line from left to right through the
vertical center of the full set of values, keeping the line as close as possible to all
values without curving it. You should use trend lines of this type in time-series
displays with caution because they can be misleading. A trend line is based only
on the values that you've included in the graph. Slightly changing the time
period that you're showing might have a significant impact on the trend line, as
you can see in this example where adding one more month to the beginning of
the line graph has caused the trend line to reverse direction.

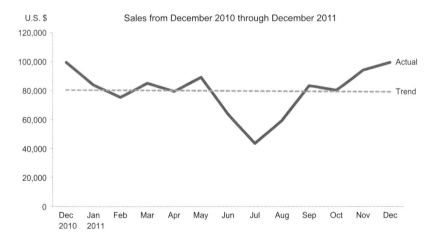

FIGURE 10.39 This trend that sloped
upward in *Figure 10.38* now slopes
downward simply because the
month of December, 2010 was
added.

Because of this potential for misrepresentation, it often makes sense to show the
overall trend of change through time by using a *moving average* (a.k.a. *running
average*). This approach softens the jaggedness that often occurs in change
through time by replacing actual values with the average of several consecutive
values. For example, a five-month moving average of monthly sales would show
for each month the average of that month's sales and the four previous months.

Here's the same sales data as above, but in this case the trend is shown as a five-month moving average.

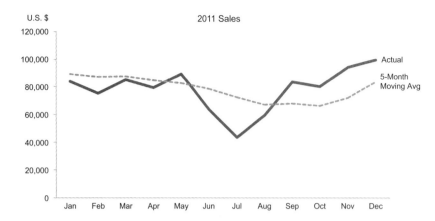

FIGURE 10.40 In this graph the overall trend is represented by a moving average.

Trend lines should always stand out as different so they aren't mistaken for actual values. In the three previous examples, I made the trend line lighter and dashed, to alert readers to its difference in meaning. It certainly isn't necessary to use dashes for a trend line although this often works well. In all examples so far, I've deemphasized the trend line relative to the line of actual values to make the latter stand out as more important. If the trend line is more important to the story that you're trying to tell in the graph, you should do the opposite to feature the trend above the actual values.

TREND LINES IN CORRELATIONS

Trend lines are often included in scatter plots to show the overall pattern of correlation. Here's a simple example:

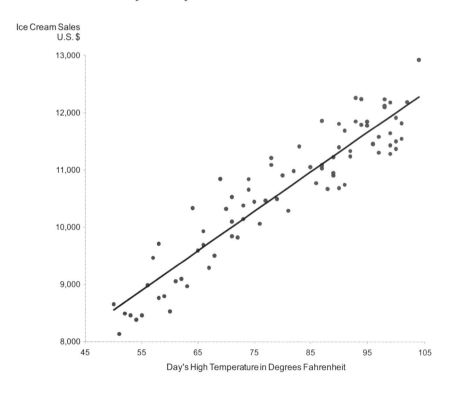

FIGURE 10.41 This scatter plot includes a trend line that highlights a correlation.

Trend lines do a great job of revealing the forest when your readers might otherwise get lost in the trees. Especially when you're dealing with a large number of erratic values, trend lines reveal what might be hard to see otherwise.

As with time-series trends, the overall patterns of correlations are frequently not linear. To do its job, a trend line should summarize the overall shape of the data. In the following scatter plot, the pattern formed by the values cannot be summarized by a straight line.

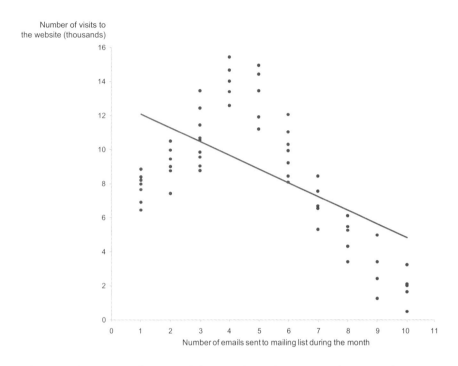

FIGURE 10.42 The trend line in this scatterplot does not match the pattern in the data.

In this case, beginning from the left the values slope upwards to a peak at 4 on the X axis and then turn downwards from there. A curved line such as the one shown below is needed to summarize the overall shape of the data.

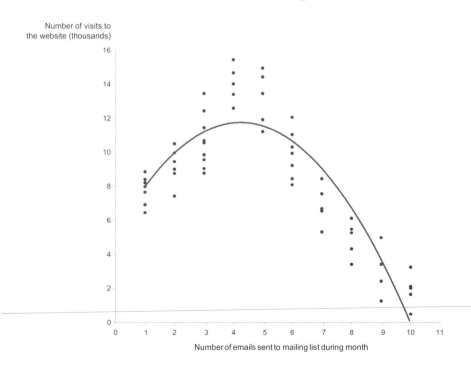

FIGURE 10.43 The trend line that describes this correlation curves to fit the data.

The story of this data set seems to be that people on the mailing list respond positively to emails by visiting the website but only up to a certain point—four emails in a month—and every additional email beyond that produces a decrease in website visitors.

The primary design practice to keep in mind for trend lines is the use of visual attributes such as hue, color intensity, line patterns, and line thicknesses to either highlight or subdue the lines, as needed.

Reference Lines

Reference lines are primarily used to provide context for the information in the form of a meaningful comparison. If you're displaying a year's worth of monthly sales revenues, you might want to include reference lines to show the previous year's lowest and highest monthly revenues, illustrated below.

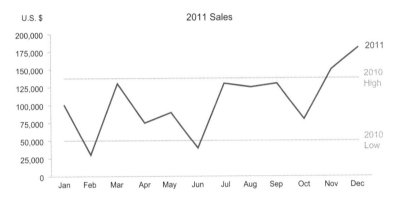

FIGURE 10.44 This graph includes reference lines to mark the highest and lowest monthly revenues during the previous year.

Reference lines are especially useful for displaying a measure of the norm, making it easy to see how values deviate from that norm. Averages (e.g., the mean or median) and standard deviations work well with reference lines. Here's an example of the same monthly sales values as in the graph above, but this time a reference line allows the viewer to compare them to average monthly sales:

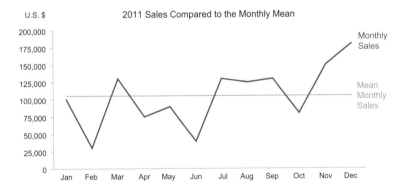

FIGURE 10.45 This graph shows a reference line that represents mean monthly sales to which each month's sales can be compared.

The next example is similar, but this time three reference lines are included to provide additional measures of the norm: one displays the mean, and two display one standard deviation above and below the mean.

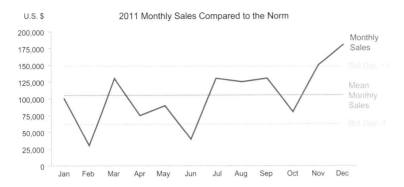

FIGURE 10.46 This graph shows reference lines that represent multiple measures of the norm.

Sometimes it works even better to mark entire reference regions. For instance, rather than using reference lines to mark standard deviations in the previous example, the entire range within one standard deviation above and below the mean could be displayed using fill color, such as in the example below.

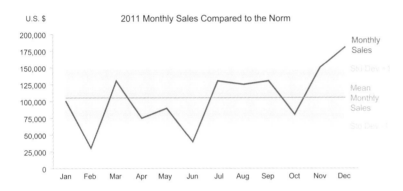

FIGURE 10.47 This graph includes reference lines in the form of a gray rectangle that marks a region of normal activity, in this case one standard deviation above and below the mean monthly sales.

If the purpose is to make values that fall outside the range of normal stand out, this does the job wonderfully, making it easy to distinguish between the normal range and everything else.

In addition to reference lines that mark particular quantitative thresholds, lines can also be used to separate categorical items into meaningful groups. In the following example, a line is used to separate products into two groups, hardware and software.

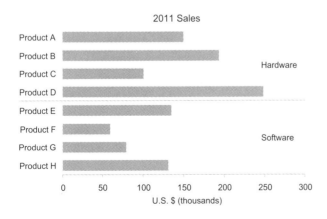

FIGURE 10.48 This graph shows a reference line that is used to mark a break point in a series of categorical subdivisions, in this case dividing products into two groups: hardware and software.

Notice that this example includes text annotations to label the two groups. We'll look more closely at annotations in a moment.

Reference lines and lines that are used to separate items into groups can be formatted in a number of ways, using various hues, styles (solid or dashed), and thicknesses, depending on the role that the lines play. As with trend lines, when you place a reference line in a line graph, you should be careful to make it look different from lines that encode the primary data. In addition, you must decide how much the line needs to stand out to do its job and then format it accordingly. In the previous example, to separate the bars into two groups, one for hardware and one for software, the line did not need to stand out very much, so it is relatively thin and light gray. If you used a reference line to mark the threshold of acceptable performance and the primary point of your story was to highlight values that exceed that threshold, you would make the line stand out to a degree that catches the reader's attention.

Annotations

Text appears in graphs in the title, the titles of and labels along axes, and labels in legends. It can also be used to annotate a graph in various ways. When words are needed to clarify meaning or to make a point, they belong in or near the graph. Regardless of its purpose or form, text should be tightly integrated with the graph but in a way that does not distract from important graphical elements. By convention, titles generally appear just above the data region of the graph, and notes generally appear just below the graph. As long as they don't disrupt perception of the visual information in the data region, notes may be placed in the data region, especially when they apply to particular data. Notes that are instructional or interpretive in nature usually don't need to appear in the data region and can be placed adjacent to the graph (below, above, or to either side), close enough that they are not overlooked. Here's an example of effective text placement:

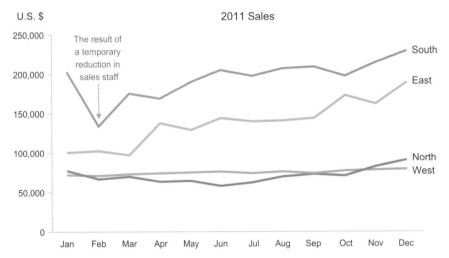

FIGURE 10.49 This graph Illustrates how annotations can be located near the point of reference without interfering with the data.

Note: The relatively even distribution of sales in the west can be attributed to the sales compensation pilot program.

In the following example, annotations point out relevant events:

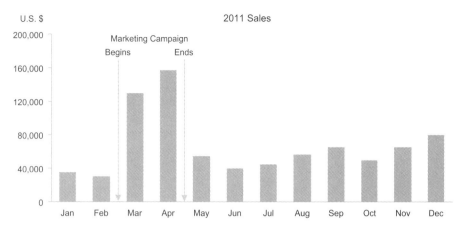

FIGURE 10.50 This graph shows annotations that identify significant events in time.

Another annotation that is often useful involves labeling values that deserve to be featured. In the next example, the highest values along each line have been labeled to draw attention to them.

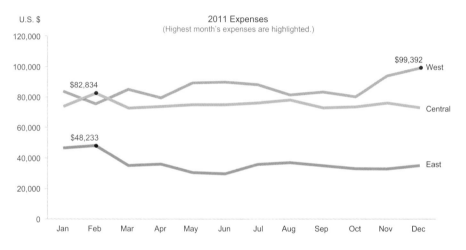

FIGURE 10.51 Particular values have been featured by labeling them.

As with reference lines, annotations should be tightly integrated in a way that complements other information without distracting from it. The prominence of annotations should be determined by their relative importance to the story that you're trying to tell. When placing annotations in the data region of a graph, it is often useful to make them inconspicuous to prevent the appearance of clutter. Reducing the intensity of text from black to gray can often accomplish this because it makes the text lighter than other data elements and thus prevents distraction, as illustrated in the example on the following page:

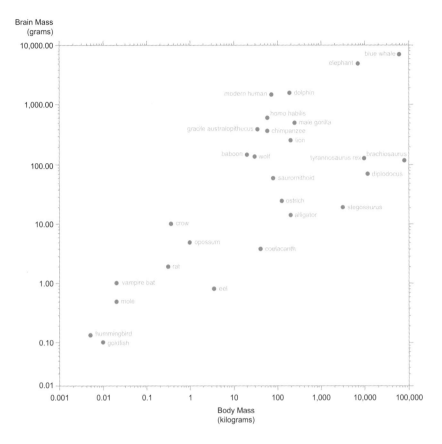

FIGURE 10.52 Values can be labeled in a way that does not distract from the values themselves. A version of this graph originally appeared in Carl Sagan's book *The Dragons of Eden: Speculations on the Evolution of Human Intelligence* (1977), Random House.

If your primary goal is to show the pattern of correlation between body mass and brain mass but you also want to allow readers to identify specific animals, this design works well. The labels can be easily read but can also be ignored when focusing on the data points because of the hue and color intensity differences between the labels and the data points.

Scales

Scales and axes are intimately related. A scale, regardless of the form it takes in the world at large (e.g., a ruler or a weight scale), measures the value of something. In graphs, scales are laid out along axes. They divide axes into increments of equal lengths and assign quantitative measures or categorical labels to those increments. Quantitative scales provide a means of assigning specific numeric values to the data objects that encode those values, based on the location of the objects along the scale line. Categorical scales associate categorical items with data objects in the same manner. The tiny lines that intersect scale lines to mark the increments are called *tick marks*; we'll cover them in the next section. Without scale lines on the axes, graphs could not tell their stories sufficiently.

Quantitative scales come in a variety of types, but graphs rarely require more than the following two:

- Linear scale
- Logarithmic scale

Along a *linear scale*—the type used in most graphs—the quantitative interval from one tick mark to the next is always the same. If the scale starts at 0, and the next tick mark as you go up the scale has a value of 10, then you know that the next tick mark after that will have a value of 20, and so on. To calculate the value of each tick mark, you simply add the value of a single interval to the value of the previous tick mark.

A *logarithmic* (abbreviated as *log*) *scale* works differently. Each value along a logarithmic scale increases by a particular rate rather than a particular amount. Another way of saying this is that each value is calculated by multiplying the previous value by a particular number, called the base. Here's an example of a log scale with a base of 10:

Actual Values	1	10	100	1,000	10,000	100,000	1,000,000
Logarithmic Values	0	1	2	3	4	5	6

FIGURE 10.53 This illustration shows a log scale with a base of 10.

What's the base 10 logarithm of 10,000? The answer is 4, since 10^4 (i.e., 10 to a power of 4, or $10 \times 10 \times 10 \times 10$) equals 10,000. The standard way of writing this is $\log_{10}(10,000) = 4$. Here's a concise definition:

> The logarithm of a number is the mathematical power to which another number, called the base, must be raised to equal that number.

You might have noticed when looking at the scatter plot that correlated body mass to brain mass in *Figure 10.52* that the scales along both axes looked like this log base 10 scale.

In a log scale, then, actual values from one interval to the next are equal to the previous value multiplied by the base. For example, in *Figure 10.53*, the actual value 1,000 is equal to the previous value, 100, multiplied by the base of 10.

If you couldn't define it before, now you can, but what good is it? When are log scales useful in graphs? Take a look at the following scatter plot, which displays the correlation between income and lifespan for various countries. The lifespan scale on the Y axis is linear, ranging from an average lifespan per person of 40 to 85 years, but the income scale in international dollars on the X axis is logarithmic, ranging from an average annual income of $100 to $100,000 dollars per year.

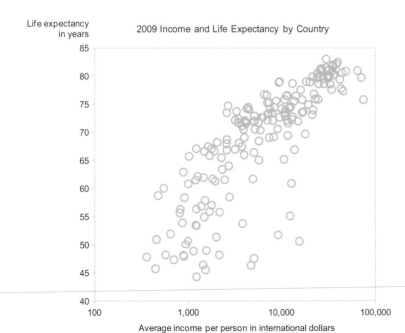

FIGURE 10.54 This scatter plot has a log scale on the X axis.

Why wouldn't this graph work well with a linear scale on the X axis? Of the 179 countries in this scatter plot, the average per-person incomes in 16 are less than $1,000, and in 76 they are less than $5,000. In a scatter plot with a linear scale, the data would look like this:

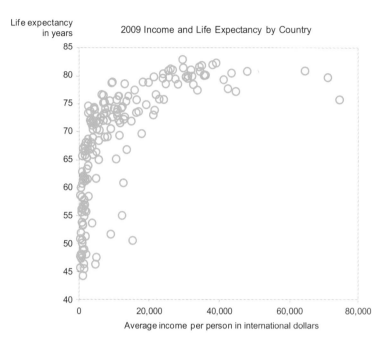

FIGURE 10.55 This scatter plot has a linear scale on the X axis.

For many purposes, this would work fine, but if you wanted to differentiate countries with low incomes, you wouldn't want them buried in a clutter of overlapping points. A log scale is a handy alternative on these occasions, but only if readers don't need to compare values based on distances between them. Here's another example of this situation:

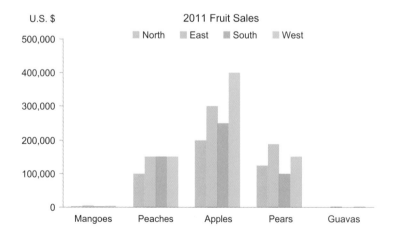

FIGURE 10.56 Several values on this graph are difficult to read because they are extremely small compared to others.

The huge difference between sales of apples at the high end and guavas and mangoes at the low end makes it difficult to interpret the low values along this linear scale. You could try to resolve this by making the graph much taller, but the result would look ridiculous. Log scales can provide an easy solution to this problem. Here are the same values, this time displayed as a dot plot with a log scale:

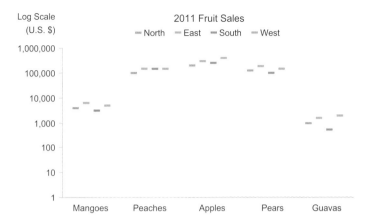

FIGURE 10.57 This graph uses a log scale to allow disparate values to be viewed together.

I substituted points for bars in this example because using the lengths of bars to represent logarithmic rather than linear values can lead to a great deal of confusion. It's hard to overcome the natural tendency to perceive a bar that is twice the length of another as representing twice the value of the other bar. Even when points are used, people can still be easily confused by a log scale, which we'll consider more carefully in a moment. For now, notice that the huge gap along the scale between the different sets of values has been dramatically reduced so that we can now see all of them more clearly and compare them more easily. Notice, however, that some care has been taken in this example to alert the reader to the fact that a log scale is being used. This is important, because readers would misinterpret differences in sales between various fruits, such as guavas and apples, if they failed to take the log scale into account and adjust their perception of differences accordingly.

With a linear scale, you know that if one value is twice as high as another, assuming the scale begins at zero, one value is twice the other. This is clearly not the case with log scales, but there is a unit of measure that will yield the same difference in value between equal differences in position along a log scale. Care to make a guess? Take a moment to think about it.

. . . .

The same distance anywhere along a log scale equals the same *percentage* or *ratio*. In the next example, I've removed two of the regions, leaving only the north and east, and have added short black vertical lines to feature the distance between sales in the north and east for each fruit.

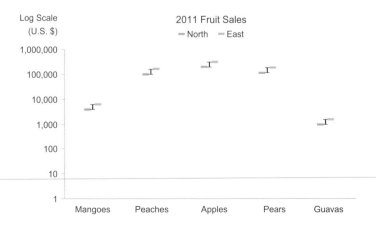

FIGURE 10.58 This example shows that, on logarithmic scales, equal distances correspond to equal percentages.

When I created this data set, I made sales values in the east precisely 50% greater than sales in the north for each fruit. For instance, the comparison of apple sales between the north and the east is $200,000 to $300,000, and for apples and guavas it is $1,000 to $1,500; in both cases a difference of 50%. The fact that the percentage difference between sales in the north and east for all fruits is the same is something that we couldn't see if the graph had a linear scale.

This feature of a log scale can be put to use, can't it? Whenever you want to compare differences in values as a ratio or percentage, log scales can do the job nicely. They function in this way because, whereas equal distances along a linear scale are equal in value, equal distances along a log scale are equal in ratio. In *Figure 10.58* on the previous page, beginning with the value of 1 at the bottom of the scale, each successive value is precisely 10 times the previous value. This is especially useful when displaying time series if you want to make it possible for your readers to compare rates of change through time. In the next example, the line graph has a linear scale, which makes it difficult to compare rates of change in sales from quarter to quarter among fruits and nuts.

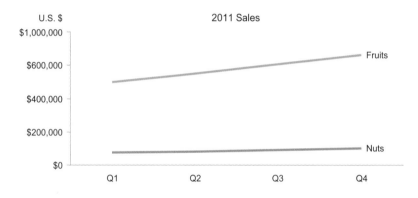

FIGURE 10.59 In this graph, time-series information is displayed using a linear scale. With a linear scale, the rate of change through time is difficult to discern.

In this next example, the same values are displayed using a log scale:

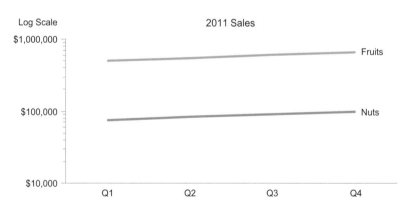

FIGURE 10.60 In this graph, the same time-series information shown in *Figure 10.59* above is displayed using a log scale to make rates of change easy to compare.

Notice that the slopes of the two lines are precisely the same. When you use lines to encode time-series values along a log scale, equal slopes indicate equal rates of change. In fact, sales of fruits and nuts both increased by exactly 10% each quarter. When the rate of change (rather than change in quantity) is the heart of your message, a log scale will serve you well. Just be sure to explain how they work to your readers because it's not at all obvious.

Despite their usefulness, the down side of log scales is the fact that they are not visually intuitive. Equal distances along the scale correspond to equal log values (ratios) but not equal linear values (amounts). Note that the minor tick marks (i.e., the smaller ones that aren't labeled) along the log scale in the last figure are not equally spaced. Each minor tick mark represents an additional increment of the value represented by the labeled tick mark below it. The first minor tick mark above the one labeled $10,000 represents $20,000 (i.e., $10,000 + $10,000). The one above that represents $30,000, and so on up the scale until you reach $100,000; above $100,000, each minor tick mark represents an additional $100,000. By displaying the minor tick marks, you can give your readers a visual clue that the scale is not linear, but this isn't enough if they aren't accustomed to log scales. If you wish to use log scales in the ways that I've described, and your readers don't already understand how to read them, you must take the time to provide a little education. You can head off inevitable confusion by including notes of explanation next to the graph where they can't be missed.

One more point to understand about log scales is that they don't have to use a base of 10. The base can be any number, but a base of 10 is the most common. In fact, some software restricts you to a base of 10. This is unfortunate, for log scales work in exactly the same way and offer exactly the same advantages with any base, and sometimes base 10 is too big to provide the best spread of values in a graph.

Base 2 is every bit as useful as base 10. With a base of 2, every major tick mark on the scale is exactly twice the actual value of the one before it (2, 4, 8, 16, 32, 64, 128, etc.). A base-2 log scale makes it easy to interpret rates of change in terms of doubling. With a time-series display in a line graph, a line that ascends upwards during an interval of time (e.g., month or year) from one major tick mark to the next indicates that the value doubled during that period. The other advantage of a base-2 log scale is that if the values you are graphing are not spread across a broad range, this scale allows you to display greater detail than a log 10 scale that jumps from 10,000 to 100,000 to 1,000,000 in single bounds.

Tick Marks

The information contained in a scale line consists solely of the tick marks, which establish the positions of values along the scale, and the labels, which assign numbers to the tick marks. The line on which the tick marks reside encodes no information itself. We'll explore the following questions about tick marks:

- How visible should tick marks be?
- Where should tick marks appear on an axis?
- When can you eliminate tick marks?
- When should you use minor tick marks?
- How many tick marks should you use?
- Which values should you indicate with tick marks?

HOW VISIBLE SHOULD TICK MARKS BE?

Tick marks and labels provide critical information because they allow us to interpret the data values in a graph, but they are not the values themselves. Consequently, they should be visually subdued in comparison to the actual data values in the graph but prominent enough to be read easily. Thin, light gray lines work well as a default.

WHERE SHOULD TICK MARKS APPEAR ON AN AXIS?

You have three options for the location of tick marks: the inner side of the axis, the outer side of the axis, or across the axis. The following example illustrates these options:

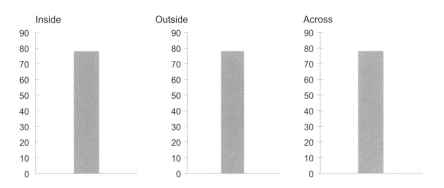

FIGURE 10.61 These examples show the three options for placing tick marks on an axis: inside, outside, and across.

As long as the tick marks are visually subdued in relation to the data values, any of these options will work, but I prefer placing them on the outside because this leaves the data region of the graph completely free of all but the values themselves, providing a clean, uncluttered backdrop.

WHEN CAN YOU ELIMINATE TICK MARKS?

Tick marks are always needed for quantitative scales, but what about categorical scales? In all but one case, tick marks are superfluous on categorical scales. The categorical labels themselves identify the locations of the values clearly enough without assistance from tick marks. Take a look at the two bar graphs below, one with categorical tick marks and one without:

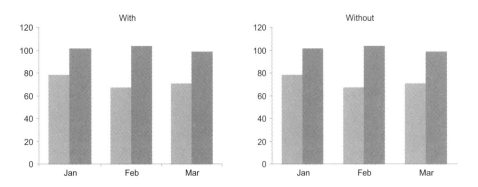

FIGURE 10.62 Tick marks appear along the categorical scale on the left but not on the right.

As you can see, tick marks along the categorical scale of a bar graph aren't needed. The same is true of graphs that encode values using points or boxes. Even though tick marks are small, if you eliminate them when they aren't needed, you'll produce cleaner, less cluttered graphs.

The one case in which tick marks can be slightly useful on a categorical scale is in line graphs that don't already mark the positions of values along the lines (e.g., by using points or grid lines). But only use tick marks when knowing the precise location of values is necessary. In the following example, notice how difficult it is to spot precisely where along the lines the value for February is positioned because the lines are relatively straight without an angle to mark the position:

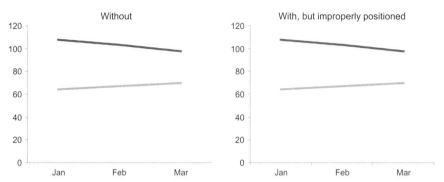

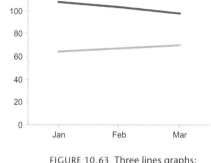

FIGURE 10.63 Three lines graphs: one without tick marks on the categorical scale, one with them but in the wrong position, and one with them in the correct position.

I included the middle example of a line graph with tick marks that are incorrectly positioned because this is a common default of software, including Excel. Even when they are correctly positioned, as on the right, tick marks are sometimes far from the lines, which makes it only slightly easier to determine the exact position of values. For this reason, if the precise location of values along the lines is really needed, it usually works best to eliminate the tick marks and either replace them with light vertical grid lines or to include points to mark values on the lines.

WHEN SHOULD YOU USE MINOR TICK MARKS?

What about minor tick marks, the ones that can be included between the labeled tick marks to indicate finer increments along the scale line? Minor tick marks suggest a level of quantitative precision that graphs just aren't meant to provide. Minor tick marks are only useful on log scales, not for the purpose of greater precision, but to alert readers that a linear scale isn't being used.

HOW MANY TICK MARKS SHOULD YOU USE?

Another consideration is how many tick marks you ought to include on the scale line. Always aim for a balance between having too many, resulting in clutter, and too few, which would make it hard for your readers to determine the values of data objects that fall between them. Here are examples of both extremes:

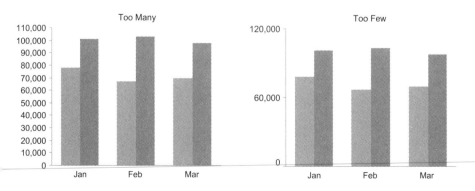

FIGURE 10.64 These are examples of graphs with too many and too few tick marks.

There is no exact number that works best in all circumstances, and the size of the graph is a factor that you must consider: the longer the scale line, the more tick marks it should contain. The following example shows a more appropriate number of tick marks than the two examples above, for the given scale length.

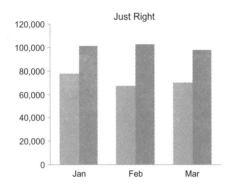

FIGURE 10.65 This graph shows an appropriate number of tick marks.

WHICH VALUES SHOULD YOU MARK WITH TICK MARKS?

Our final consideration regarding tick marks is the choice of values that they designate. It generally works best to use nice round numbers to which your readers can easily relate. Avoid irregular values like the following:

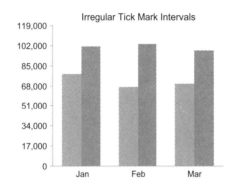

FIGURE 10.66 This graph shows tick marks that label values at peculiar increments, making them difficult to interpret.

If you use numbers like those above, you'll make your readers work extra hard and lead them to wonder why the increments are so peculiar, searching for meaning where none exists.

Grid Lines

From the early days of quantitative graphs, one of the primary purposes of grid lines was to make it easier to draw graphs. The lines were preprinted on graph paper as guides to assist the placing and sizing of objects. Now that almost all graphs are computer generated, this purpose is obsolete. Today, grid lines can be used to assist graph perception in three ways:

- Ease look-up of values
- Ease comparison of values
- Ease perception and comparison of localized patterns

In all cases, grid lines are support components, assisting in the visual perception of the data values themselves. Thus, grid lines should always be visually subdued. Thin, light lines (e.g., light gray) usually work best. Dark or heavy grid lines, such as those in the example below, should always be avoided because they disrupt perception of the data and produce annoying clutter.

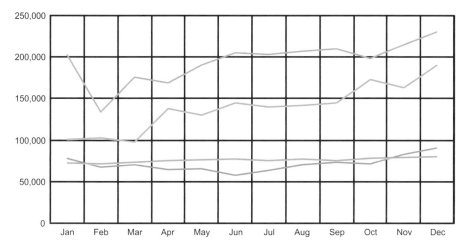

FIGURE 10.67 This graph includes dark, heavy grid lines, which you should always avoid.

Notice how much easier it is to focus on the lines of data in the example below.

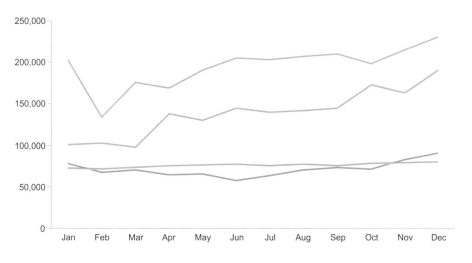

FIGURE 10.68 Without the dark grid lines, it is much easier to focus on the data.

ENHANCE THE LOOK-UP OF VALUES

The most common but least necessary use of grid lines is to assist readers in looking up values. We established previously that tables work much better than graphs for looking up precise values. Because graphs are used primarily to present the shape of the data, precise values usually aren't necessary. Nevertheless, grid lines do help when tick marks alone do not provide the level of precision that your readers would find useful for interpreting data values in graphs. This situation is most common with graphs that are especially wide, which makes it difficult to assign values to data objects that reside far from the scale line. In the following example, subtle grid lines provide all the look-up assistance that is necessary.

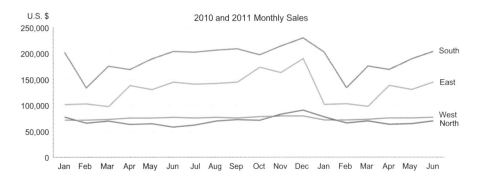

FIGURE 10.69 This graph uses subtle grid lines to enhance look-up.

ENHANCE THE COMPARISON OF VALUES

When perception of subtle differences is necessary, grid lines help by dividing the data region into smaller units, which makes otherwise subtle differences in size and location obvious. Grid lines allow us to compare differences in the size and position of objects relative to a much smaller space than the entire data region of the graph, which significantly increases the percentage differences between objects, making these differences easier to see. The following examples illustrate this point:

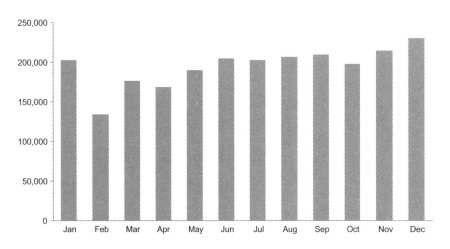

FIGURE 10.70 This graph demonstrates how difficult it is to perceive differences in the size and location of data objects when those differences are relatively small.

It's difficult to assess differences between bars that are close in length, such as January, June, and August. However, with the addition of light horizontal grid lines in the example below, these subtle differences become much easier to discern:

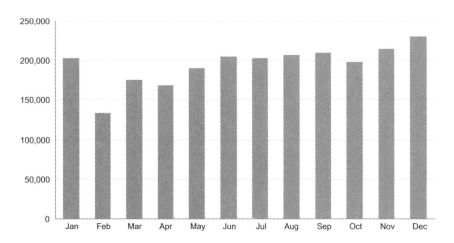

FIGURE 10.71 This graph uses grid lines to make it easier to perceive differences in the lengths of the bars.

ENHANCE PERCEPTION AND COMPARISON OF LOCALIZED PATTERNS

Graphs do a marvelous job of revealing the overall shape of an entire collection of values. And if you need to enhance perception of values in a subsection of the graph, grid lines can provide useful assistance. Imagine that you're examining the correlation between people's heights and salaries, which you've broken into two regional groups: West and East. For the moment, you want to compare values in the West to those in the East that fall from 65 to 70 inches in height and from $50,000 to $60,000. The grid lines in the graphs below would make this easy, but you'd struggle without them.

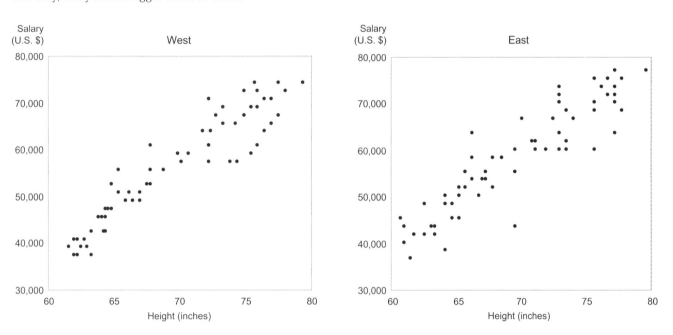

Lines that don't correspond to every value along scale (technically not grid lines) can also be used in a scatter plot to delineate meaningful sections of data. For instance, it is sometimes useful to divide a scatter plot into quadrants, as in the following example:

FIGURE 10.72 It would be difficult to isolate and compare regions of data in these scatterplots without the grid lines.

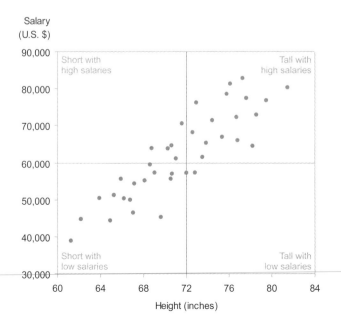

FIGURE 10.73 This scatter plot uses two intersecting lines to evenly divide the data region into quadrants.

The use of lines to divide scatter plots into quadrants can bring features of the data to light that might otherwise be missed. The previous example clearly reveals, with its nearly empty lower right quadrant, that tall employees rarely receive low salaries.

Legends

Legends are the tiny tables that associate categorical items with the visual attributes that encode them (e.g., hue, color intensity, or point shape). We'll look briefly at the following issues:

- When can you eliminate legends?
- Where should legends appear on the graph?
- How visible should legends be?
- Should legends have borders?
- How should you arrange labels in legends?

WHEN CAN YOU ELIMINATE LEGENDS?

Whenever categorical items are encoded in a graph, they must be labeled. If they appear on a categorical scale along an axis, they are labeled along the axis. However, if they are encoded using some other visual attribute, such as hue, they must be labeled in some other way. Legends are the conventional way to label these categorical items, but one circumstance lends itself to an alternate form of labeling. The graph below gives a hint. Take a look at it and see whether you can determine a different way to label the lines.

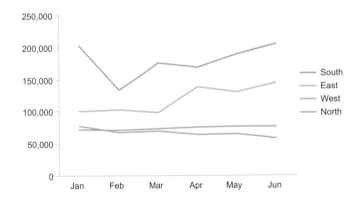

FIGURE 10.74 This graph includes a legend that could be replaced using a different means of labeling the categorical items of South, East, West, and North.

It's fairly easy to recognize in this example that the legend could be replaced by labeling the lines directly as you've seen me do on several occasions. Here's the improved version:

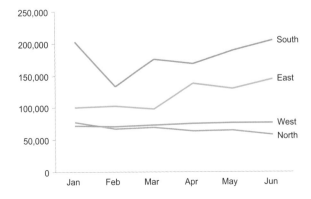

FIGURE 10.75 In this graph categorical items are labeled directly, next to the lines that encode them, rather than separately in a legend.

If you are producing graphs using software that doesn't offer the option of placing labels right next to the data sets, you can often type the labels in yourself.

The answer to the question "When can you eliminate legends?" is "Whenever the data items that need labels are grouped together so that you can place a label right next to each set." When lines are used to encode sets of values, each line acts as a grouping mechanism, making it easy to place a label right next to the line. However, as you can see in the next example, bars don't work this way:

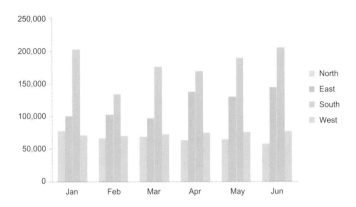

FIGURE 10.76 This is a graph that must use a legend to label the categorical items because the values are encoded as bars.

You certainly wouldn't want to repeat the same labels over and over again for each bar.

WHERE SHOULD YOU PLACE LEGENDS?

Legends are generally placed outside of the data region, but this needn't be a rigid rule. If you can place the legend inside the data region without getting in the way of the data values or interfering with perception of their shape, there is no reason why you shouldn't put it there. In fact, the closer the legend is to the data values, the easier it is to read the graph. Legends may be placed anywhere they fit and don't interfere with other more important components of a graph.

HOW VISIBLE SHOULD LEGENDS BE?

Legends provide the means to interpret categorical data, which is critical information, but they are not the actual data themselves. Therefore, although legends should be clearly visible in the graph, they should be somewhat less prominent than the actual data. You want your readers' eyes to be drawn predominantly to the data region of the graph.

SHOULD LEGENDS HAVE BORDERS?

You may have noticed that few of the legends that appear in examples in this book have borders around them. This is an intentional omission. Based on your understanding of visual perception, why do you think this is? Look at the two graphs below to see the difference:

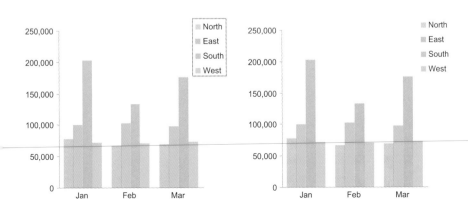

FIGURE 10.77 These two graphs are the same except that in one a border surrounds the legend and in the other the legend has no border.

First, the border around the legend does not add any meaning to the graph, and the legend doesn't need a border in order to be clearly distinct from other parts of the graph. Second, enclosing the legend with a border draws undesirable attention to it. The legend plays only a supporting role, so it isn't where you want your readers' eyes to be drawn. And finally, the border serves as a container, perceptually separating the legend information from rest of the graph.

When a legend is placed near another graph component and these two components must appear distinct from one another, something is needed to visually distinguish the legend from that other component; in this case, a border is useful. Be sure to make the border light, just visible enough to set the legend apart.

HOW SHOULD YOU ARRANGE LABELS IN LEGENDS?

You are probably accustomed to seeing the labels in legends arranged vertically rather than horizontally, but either arrangement can work fine. It really depends on the space you're working with and how the data is encoded. In the following example, the legend is arranged horizontally below the title, which prevents the graph from being wider than necessary.

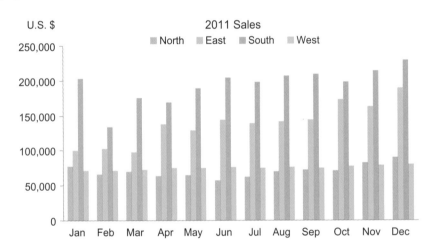

FIGURE 10.78 In this graph the labels in the legend are arranged horizontally.

This legend's horizontal arrangement of labels and position above the data region save horizontal space in the graph. The labels are also arranged in the order of the corresponding bars, which improves the graph's ease of use.

HOW MANY ITEMS CAN A LEGEND CONTAIN?

There is no conceptual limit to the number of categorical items you can include in a single graph, but there is definitely a practical limit. Too many visually distinct categorical items cannot be decoded without laborious effort because, as you learned in Chapter 5, *Visual Perception and Graphical Communication*, we can only hold the meanings of a few distinctions in working memory at one time. We've all seen graphs like the following:

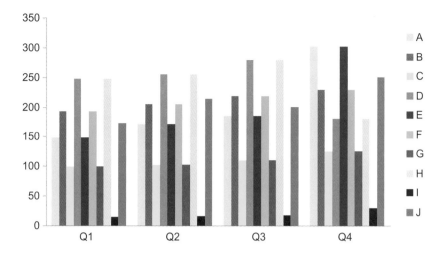

FIGURE 10.79 This graph has too many categorical subdivisions.

In fact, many of us have created similar graphical monstrosities. Not only is it impossible to keep the meaning of each color in mind, the graph also appears horribly cluttered. Clutter is visually exhausting, discouraging efforts to read the graph.

Non-Data Component Design

Axis lines are the only non-data components that are routinely useful in graphs. Axes give graphs dimension and serve as a container for the data. A single axis, either vertical or horizontal, produces a 1-D graph. Two perpendicular axes, one vertical and one horizontal, produce a 2-D graph, which is by far the most common type. The space that is defined by the axes where the data values are plotted (i.e., where the points, lines, or bars reside) is called the data region or the plot area.

Axes

When you design a graph, you need to ask two questions regarding the axes:

- Should the graph include one, two, three, or four axis lines?
- What ratio of the lengths of the X and Y axes works best?

SHOULD THE GRAPH INCLUDE ONE, TWO, THREE, OR FOUR AXIS LINES?
2-D graphs usually include one vertical axis on the left and one horizontal axis along the bottom. Under most circumstances, this convention is all that's needed. It is sometimes useful, however, to change the position of the axes. In the example that appears on the next page, you will see conventional axis placement as well as the alternatives:

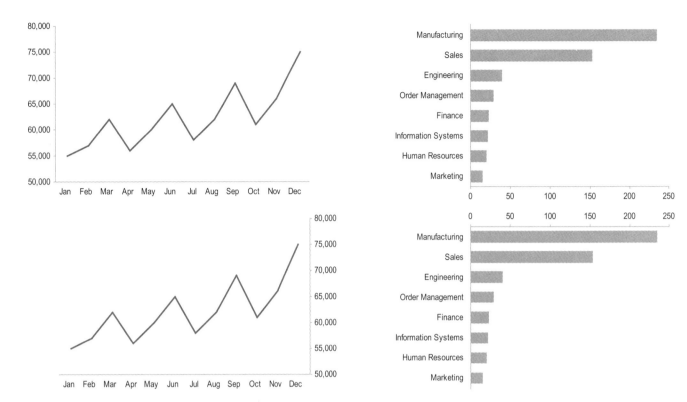

FIGURE 10.80 The upper graphs position the axes and scales conventionally and those below show the alternatives.

Looking at the lower line graph, can you think of an occasion when placing the Y axis on the right would make sense?

· · · · · · ·

One obvious benefit is the fact that it is now easier to decode values along the line that appear on the right because the scale is closer to them. Did you also notice how strongly your eyes are drawn to the right half of the graph by the fact that the Y axis and scale are there. One occasion when this switch is useful is when you want to feature values on the right (e.g., the months in the final quarter of the year) more than those on the left. The mere presence of visual content on the right that is missing on the left draws the viewer's attention to the right. Switching the position of the axis from left to right or from bottom to top is a useful way to direct your reader's attention to the opposite region of the graph without adding anything to the graph to accomplish this.

Regardless of position, are there times when more or less than two axis lines are useful alternatives? In the following illustration, the conventional configuration of two axis lines is shown along with three alternatives:

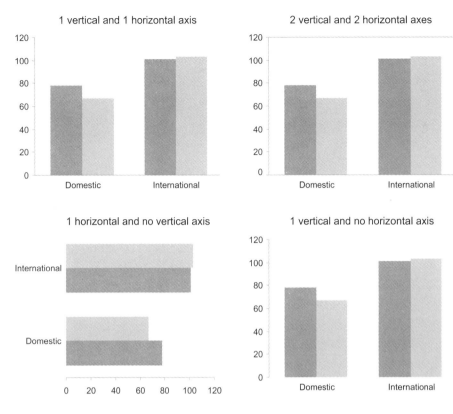

FIGURE 10.81 These graphs illustrate various axis configurations.

Two perpendicular axes tracing only two sides of the rectangular data region sufficiently define the space in most cases. Including two pairs of vertical and two pairs of horizontal axes to form a complete border around the data region is mostly useful when the data region must be clearly separated from surrounding content because the surroundings would otherwise compete too forcefully for attention. An axis may be left off without adverse affect when it hosts a categorical scale, and the values are encoded as horizontal bars. Because bars begin at the axis, their edges trace the line that the axis would otherwise display, which adequately delineates the data region. This works fine for horizontal bars, but I find that, without a base, vertical bars appear to float in space, as in the graph on the bottom right.

WHAT RATIO OF THE LENGTHS OF THE X AND Y AXES WORKS BEST?

The ratio of the length of the horizontal axis to the length of the vertical axis, or stated differently, the ratio of the data region's width to its height, is called the *aspect ratio*. It is calculated as width divided by height. The aspect ratio of a graph's data region greatly influences perception of the data. On the next page are a few examples of aspect ratios that vary from 2 to 0.5.

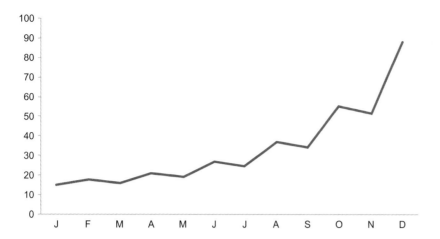

FIGURE 10.82 This graph has an aspect ratio of 2 to 1, or 2.

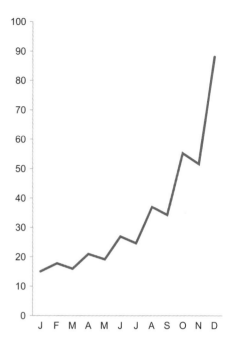

FIGURE 10.85 This graph has an aspect ratio of 1 to 1.5, or 0.67.

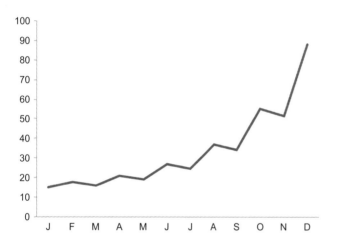

FIGURE 10.83 This graph has an aspect ratio of 1.5 to 1, or 1.5.

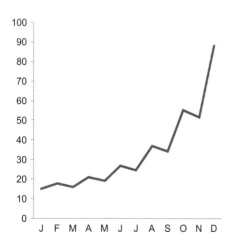

FIGURE 10.84 This graph has an aspect ratio of 1 to 1, or 1.

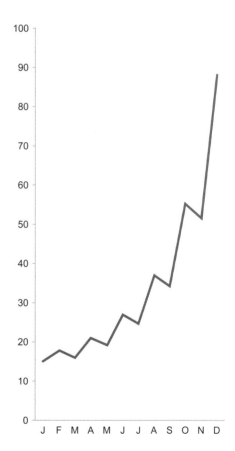

FIGURE 10.86 This graph has an aspect ratio of 1 to 2, or 0.5.

All of these graphs contain the same data; only their aspect ratios differ. When lines are used to encode values, as the aspect ratio increases, so does the appearance of the degree of change. A degree of change that already looks impressive with an aspect ratio of 2.0 looks like a blast-off at Cape Canaveral with an aspect ratio of 0.5. Both are accurate, both display precisely the same numbers, but they certainly differ in how they would be perceived.

There is no single aspect ratio that is always best. There are, however, two design practices that you should keep in mind. The first is that you should never manipulate the aspect ratio to intentionally exaggerate or downplay the degree of change. If your graphs usually appear wider than they are tall, suddenly making one taller than it is wide to convince your readers that sales are going through the roof would qualify as manipulation. The other general practice is to stick to the convention of making your graphs that display time series wider than they are tall. Emphasizing the horizontal rather than the vertical generally makes time series a bit easier to read and more in line with what people are accustomed to seeing.

Data Region

Apart from determining the data region's aspect ratio, we must design it to be a clean background in front of which the data can be easily seen. Readers' eyes should be drawn to the data. This is accomplished by making the points, bars, lines, or boxes visually prominent, not by energizing the background with vibrant color, a perception-skewing color gradient, or a silly image.

Data objects will stand out well against a light background unless they are excessively light themselves. A white data region is usually the best background, but there are times when other light colors, such as light gray or yellow, are useful, when a little extra is needed to draw readers' eyes to the data region because other content on the page or screen competes for attention. On such occasions subtle light fill color usually does the trick. Here are a few examples of fill colors that work:

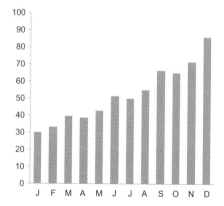 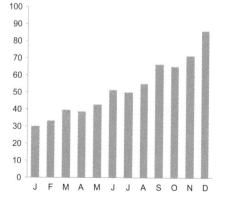 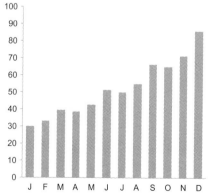

FIGURE 10.87 These examples show data region fill colors that provide an appropriate background for the data.

Another way to subtly highlight the data region is to use fill color for the graph except in the data region. By leaving the fill color of the data region white and coloring the surrounding areas of the graph with a light color, the white of the data region stands out in contrast, as in the following examples:

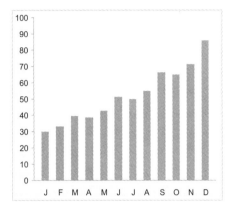 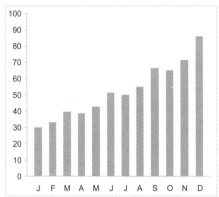

FIGURE 10.88 These graphs highlight the data region by using a subtle fill color for the surrounding areas.

When the data region needs an extra visual boost, the other visual attributes of the graph, especially the colors of the data objects, determine which of the two methods will work best, a fill color for the data region or for the surrounding areas of the graph. For instance, if your graph contains several sets of bars, and the color of one is light, the use of a subtle fill color in the data region might help them stand out more clearly than they would against white.

Gradients of color have their appeal, I suppose, but when they appear in the background of a graph, they skew perception of the data. Notice in the following example how the lines change in appearance depending on the background color.

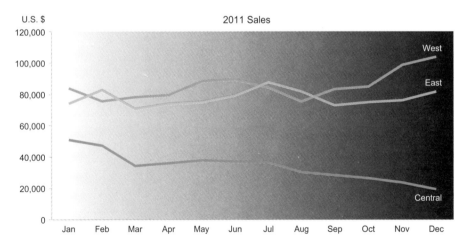

FIGURE 10.89 Color gradients in the background of the data region skew perception of the data.

Avoid gratuitous gradients and by all means resist the temptation to decorate you graphs with photographs or clip art.

FIGURE 10.90 Images in the background make it difficult to focus on the data.

An example like this might work fine in a magazine ad if you care little about readers' perception of the data, but they don't belong in the background if your objective is to tell a story contained in data. I'm not saying that photos and other images are never useful when presenting data. An appropriate photo can make a story real to people and touch their hearts in powerful way. Using a photo in conjunction with a graph in a way that complements the data, such as by placing the photo and the graph near one another, can tell a richer story without undermining the integrity of the data.

Summary at a Glance

Component	Practices

Points	• When sets of points cannot be clearly distinguished, correct by:
	  • Enlarging the points
	  • Selecting objects that are more visually distinct
	• When points overlap such that some are obscured, correct by:
	  • Enlarging the graph and/or reducing the size of the points
	  • Removing the fill colors

Bars	• Use horizontal bars when their categorical labels bars won't fit side by side.
	• Never use horizontal bars for time-series values.
	• Proximity
	  • Set the width of white space separating bars that are labeled along the axis equal to the width of the bars, plus or minus 50%.
	  • Do not include white space between bars that are differentiated by a legend.
	  • Do not overlap bars.

Component	Practices

Bars *(continued)*	• Fills
	  ○ Avoid the use of fill patterns (e.g., horizontal, vertical, or diagonal lines).
	  ○ Use fill colors that are clearly distinct.
	  ○ Use fill colors that are fairly balanced in intensity for data sets that are equal in importance.
	  ○ Use fill colors that are more intense than others to highlight particular values.
	
	• Only place borders around bars when one of the two following conditions exists:
	  ○ The fill color of the bars is not distinct against its background, in which case you can use a subtle border (e.g., gray).
	  ○ You wish to highlight one or more bars compared to the rest.
	
	• Always start bars at a baseline of zero.

Lines	• Distinguish lines using different hues whenever possible. • Include points on lines only when values for the same point in time on different lines must be precisely compared.
Boxes	• Follow the principles for bar design, except when box plots are connected with a line to show change through time, which might require greater distance between boxes.
Combinations	• Use boxes and lines for distributions through time. • Use bars and lines in the form of Pareto charts for featuring the contribution of the largest portions of the whole. • Use bars and points for uncluttered comparisons.
Trend Lines	• In most cases, use moving averages rather than straight lines of best fit to show the overall nature of change through time. • Only use linear trend lines (straight lines of best fit) in a scatter plot when the shape of the data is linear rather than curved.
Reference Lines	• Use reference lines to mark meaningful thresholds and regions, especially for measures of the norm.
Annotations	• Use text to feature and comment on values directly when doing so is important to the story.
Log Scales	• Use log scales to reduce the visual difference between quantitative data sets with significantly different values so they can be clearly displayed together. • Use log scales to compare differences in value as percentages.
Tick Marks	• Mute tick marks in comparison to the data objects. • Use tick marks with quantitative scales but not with categorical scales, except in line graphs when slightly more precision is needed. • Aim for a balance between including so many tick marks that the scale looks cluttered and using so few that your readers have difficulty determining the values of data objects that fall between them. • Avoid using tick marks to denote values at odd intervals.

Component	Practices
Grid Lines	• Thin, light grid lines may be used in graphs for the following purposes: • Ease look-up of values • Ease comparison of values • Ease perception and comparison of localized patterns
Legends	• Use legends for categorical labels when the labels are not associated with a categorical scale along an axis and cannot be directly associated with the data objects. • Place legends as close as possible to objects they label without interfering with other data. • Render legends less prominent than the data objects they label. • Use borders around legends only when necessary to separate legends from other information.
Axes	• Don't manipulate the aspect ratio to distort perception of the values.
Data Regions	• Keep the background clean and light.

11 DISPLAYING MANY VARIABLES AT ONCE

Graphs can be used to tell complex stories. When designed well, graphs can combine a host of data spread across multiple variables to make a complex message accessible. When designed poorly, graphs can bury even a simple message in a cloud of visual confusion. Excellent graph design is much like excellent cooking. With a clear vision of the end result and an intimate knowledge of the ingredients, you can create something that nourishes and inspires.

We often need to tell stories that involve more than we can clearly express in a single graph. To communicate effectively, we sometimes need to help our readers see the information from more than one perspective. A single graph can sometimes be used to tell a complex story elegantly, but frequently a single graph won't do. This chapter focuses on design strategies that address the presentation of many variables in two useful ways: combining multiple units of measure in a single graph and combining multiple graphs in a series.

Combining Multiple Units of Measure

It is often easy to combine multiple quantitative variables in a single graph when they all use the same unit of measure. For instance, in a single time-series graph as shown below, you could easily display revenues, expenses, and profits because they are all expressed in U.S. dollars.

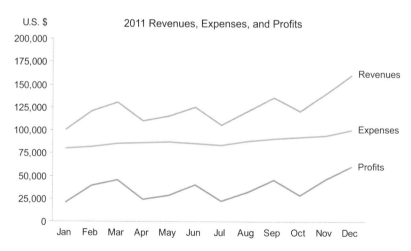

FIGURE 11.1 This graph displays multiple sets of quantitative data, all sharing the same unit of measure.

This does the job unless the values of different variables differ by a large amount, which might cause lines with low values to look relatively flat, thus making it difficult to discern their patterns. But what do you do to combine related sets of quantitative data so they can be compared when they're expressed as different units of measure? A typical example involves time-series sales information consisting of revenue in dollars and order volume as a count. One solution is to create two graphs and place them close to one another in a manner that makes comparisons easy, as illustrated below:

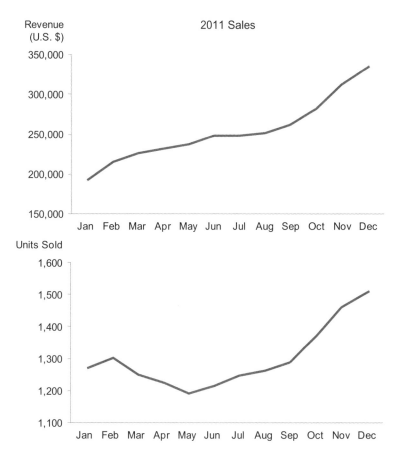

FIGURE 11.2 Arranging these two graphs in this manner allows easy comparison of patterns of change through time.

This is usually the best solution. You can, however, combine two units of measure in a single graph by using two quantitative scales as shown below:

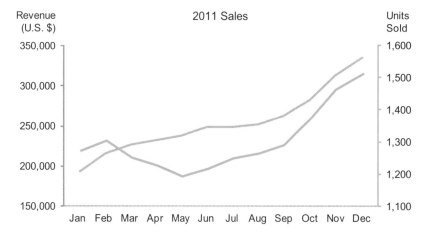

FIGURE 11.3 This single graph displays two quantitative scales on the Y axis: one on the left and one on the right.

In this configuration, even though they use different units of measure, the two related measures can be compared quite easily—perhaps too easily. Although each line is associated with a different scale, we are, by displaying them in a single graph, tempting people to do something they shouldn't: to compare the magnitudes of values on one line to those on the other. Because the scales of the two lines are different, this comparison is meaningless.

Looking again at the previous graph, do you find that your eyes are drawn to a particular spot? If you're like most people, the place where the two lines intersect catches your attention. In a normal line graph with a single quantitative scale, points of intersection actually mean something: that one set of values exceeded the other during that period of time. However, in a graph with two quantitative scales such as the one above, points of intersection are meaningless but the configuration of the graph features the intersection. This is a problem. Graphs with two quantitative scales can easily confuse and even mislead your readers. For this reason, unless you are certain that your readers are comfortable with dual-scaled graphs, it is best to avoid them.

Combining Graphs in a Series of Small Multiples

You can only squeeze so many sets of data into a single graph. It is this limitation of 2-D graphs that often tempts people to use 3-D graphs, with disappointing results. However, there is a solution that extends the number of data sets that can be displayed: multiple graphs arranged together as a series. Edward Tufte refers to the arrangement of related graphs that we'll examine in this section as *small multiples* and others call it a *trellis chart*. Tufte explains that "Small multiples resemble the frames of a movie: a series of graphics, showing the same combination of variables, indexed by changes in another variable."[1]

1. Edward R. Tufte (2001) *The Visual Display of Quantitative Information,* Second Edition. Graphics Press, page 170.

Let's walk through a typical scenario. Imagine that you need to display sales—both bookings and billings separately—by sales region for last month. Here's what you have so far:

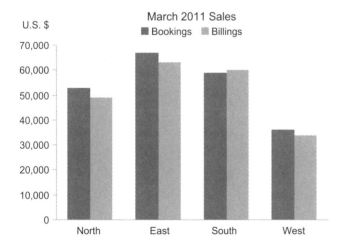

FIGURE 11.4 This graph displays two sets of quantitative values (bookings and billings) and one set of categorical items (sales regions).

Simple, so far. Now, if you need to show another set of related quantitative values (e.g., profits), you could do so easily by adding another set of bars. But what if you need to show sales not only by sales regions but also by sales channels (e.g., direct, distributor, and reseller sales)? This requires the addition of another set of categorical items. Adding a third axis would result in utter visual confusion, so this option is out. What can you do?

The answer involves multiple graphs arranged closely together so they can be easily compared. Here's a simple example that solves our problem:

FIGURE 11.5 This is a series of related graphs that displays two sets of quantitative values (bookings and billings) by two sets of categorical items (sales regions and channels).

Even though this involves three separate graphs, the nature of the arrangement allows them to be examined as one composite graph with three sets of axes. This particular series happens to consist of three graphs because there are three sales channels, but it could consist of as many graphs as you could fit together on a page or screen. As the number of graphs grows, the trick is to reduce their individual size enough to allow them to be seen together. You can arrange the graphs horizontally (side by side), vertically (one above the next), or in both directions to produce a matrix of graphs arranged in multiple columns and rows.

Graphs in a series like this should be consistent in design, varying in that each contains a different subset of data. In the example above, the graphs vary by sales channel. Because all of the graphs share a common quantitative scale, they are easy to compare.

When graphs are displayed as a series, redundant labels can often be eliminated. Notice that labels for the four regions only appear in the left-most graph and that the legend that labels billings and bookings only appears on the right. Because the graphs are close to one another, these labels don't need to be repeated in each graph. Repeating them would result in redundant and therefore unnecessary data ink.

To learn how to effectively design related graphs in a series, we'll examine the following topics:

- Consistency
- Arrangement
- Sequence
- Grid lines

Consistency

It is important when designing a series of small multiples to maintain consistency among them. Consistency is required for comparison. Consistency also contributes a great deal to efficient interpretation. Your readers only need to learn how the first graph works and can then quickly apply that knowledge to each graph in the entire series because all of the graphs work in precisely the same way.

When you design a series of small multiples, make sure that every visual characteristic of each graph within the series is the same. This includes the aspect ratio of the axes, the colors used to encode the data, the font used for text, and so on. Any difference will slow your readers down and induce them to search for meaning in that difference.

Pay particular attention to the scales along both axes. Many software products that generate graphs automatically adjust quantitative scales to fit the range of values. When graphs are combined in series of small multiples, this is a significant problem, because the graphs can only be compared accurately if their scales are the same. Look at what happens to our previous example when we allow the software to adjust the quantitative scale:

FIGURE 11.6 This example demonstrates the problem that results when quantitative scales vary in a series of related graphs.

As you can see, the differences in sales between the three sales channels appear smaller because of the variations in scale.

Make sure that the categorical scale also remains consistent with the same items in the same order (e.g., always West, South, East, and North) and the same full set of items in each graph even when a value is zero or null. Here's the same series again, but this time there are no distributor sales in the south. Rather than displaying a value of zero, the south region was excluded from the graph entirely, resulting in a confusing inconsistency:

FIGURE 11.7 This series of related graphs exhibits an inconsistency in the sets of sales regions: the south region is missing in the middle graph of distributor sales.

Arrangement

The best arrangement for a series of small multiples, whether horizontally side by side, vertically one on top of the other, or both in the form of a matrix, depends primarily on your answer to the question "Which items do you want to make easiest for your readers to compare?" Take a look at the next example to see which values are easy to compare and which are harder.

FIGURE 11.8 This arrangement of small multiples works best for comparing regions within a given sales channel and also works well for comparing values across sales channels.

This arrangement makes it easiest to compare the magnitudes of regional values within each sales channel because they are in the same graph, slightly less easy to compare values for all regions and sales channels because they are spread among three separate graphs, and least easy to compare values for a specific region in all sales channels because they are spread among three separate graphs and you must focus on one region while trying to ignore the other three.

If you wanted to make it easier to isolate particular regions for comparison across all sales channels merely by scanning across a given row, the following arrangement would support that well, but it would be harder to compare these values precisely because their quantitative scales are not aligned but are side by side.

FIGURE 11.9 This arrangement of small multiples makes it easier to focus on a particular region across all sales channels.

If you wanted to make it as easy as possible to compare sales performance for a given region across all three sales channels, then it would make sense to place the sales channels along the Y axis and create four graphs, one for each region, as follows:

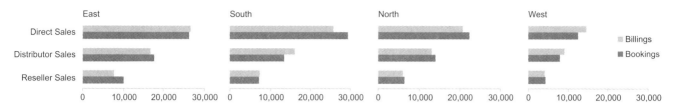

Every arrangement has its strengths and weaknesses. The best arrangement depends entirely on which comparisons your readers will most need to make. If a single arrangement won't support most of what they need to do, then multiple arrangements would be useful. Just as one graph won't always give readers everything they need, one arrangement of small multiples won't always suffice.

When a single column or row can't display all the small multiples in a series, you can arrange them as a matrix. A matrix of small multiples allows you to squeeze many onto a single page or screen. Here's an example with six scatter plots.

FIGURE 11.10 This arrangement of small multiples makes it easiest to compare sales channels in a specific region.

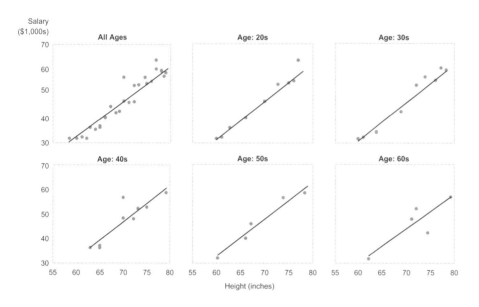

FIGURE 11.11 This series of scatter plots is arranged as a matrix. Employees' heights and salaries are correlated along the X and Y axes, and their ages are grouped into separate graphs.

As you can see, many more than six graphs could have been squeezed onto a single page or screen.

Any type of graph can be arranged in a matrix of small multiples, but this arrangement excels for the display of scatter plots, which can incorporate an extremely large amount of information for simultaneous viewing and comparative pattern detection. This allows you to display correlations among three sets of variables: two associated with the X and Y axes and a third per graph. Even when you're not using scatter plots, if you can't fit the entire series across or down a single page or screen, a matrix arrangement is a good alternative. For instance, having to scan one graph at a time across five rows of bar graphs on a single page or screen is a negligible disadvantage compared to our inability to remember them if they were spread across multiple pages or screens. Notice how nicely this works:

FIGURE 11.12 This series of small multiples has been arranged in a matrix even though it includes a categorical scale on one of the axes.

Sequence

Just like the sequencing of categorical items in a single graph, the sequencing of small multiples can contribute a great deal to their effectiveness, especially to your readers' ability to see meaningful patterns and to compare values that appear in different graphs. Take a look at the same series of small multiples that appears in *Figure 11.12* arranged alphabetically rather than from the department with the highest expenses to the one with the lowest.

FIGURE 11.13 This series of small multiples is sequenced alphabetically by department name.

It is now more difficult to compare values among the various graphs because sets of values that are close in size are no longer near one another.

The category that varies from graph to graph in a series of small multiples is called the *index variable*. Items in the index variable sometimes have a proper order. This order can help us determine how to sequence the graphs. For instance, if each of the graphs represents a different year, then you will usually want to sequence the graphs chronologically. When a meaningful order is built into the index variable and there is no overriding reason to sequence the graphs in a different order, you should sequence them accordingly. When there is no intrinsic order built into the index variable, you should sequence the graphs according to their quantitative values from low to high or high to low. Imagine that you have a series of graphs displaying sales data, and the index variable is product. Two ways of sequencing the graphs appear on the following page:

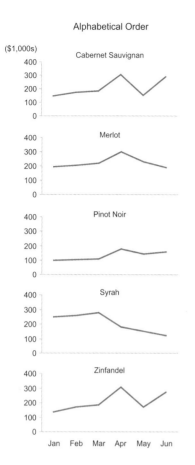

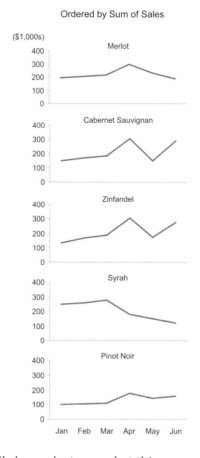

FIGURE 11.14 Two ways of sequencing a series of small multiples.

The series on the left is sequenced alphabetically by product name, but this arrangement isn't helpful, is it? The only purpose of alphabetical order is to make it easy to find individual items in a long list. With a small collection of graphs such as this, no one would need an alphabetical arrangement to find a particular product. It is much more useful to sequence these graphs by sales amounts as on the right, which makes their relative performance clear. When sequencing graphs in rank order by value, you must decide whether to base the sequence on an aggregation of all values in each graph (usually the sum or mean) or on a particular value (e.g., the final month of June). Choose the one that is most relevant to your message. If each graph includes more than one set of values, such as bookings and billings, base the sequence on the set of values that is most significant to your message.

Sums and means aren't always the appropriate values on which to base your sequence. The best choice depends on the relationship you're displaying, the values you're using to display that relationship, and the essential point you're trying to make. If you were displaying correlations using a matrix of scatter plots, it might be appropriate to rank the graphs based on the linear correlation coefficient. If you were displaying frequency distributions, the spread or median might be the best value to use for sequencing the graphs. No matter what value you use, the objective is to sequence the graphs to reveal most clearly the relationships that are central to your message.

Unless the order in which you've sequenced the graphs in a series of small multiples is obvious, it's helpful to add a note to make clear what the order is.

Rules and Grid Lines

Rules or grid lines can be used to delineate graphs in a series of small multiples but are rarely needed. They are only useful when one of the following circumstances exists:

- White space alone is not sufficient to delineate the graphs because of space constraints.
- The graphs are arranged in a matrix, and white space alone cannot be used to direct your readers to scan the graphs in a particular sequence, either across the rows or down the columns.

The following examples illustrate how rules can be used to direct readers to scan primarily in a horizontal direction.

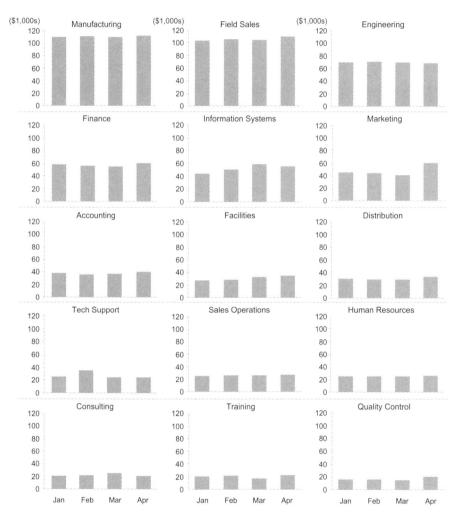

FIGURE 11.15 This matrix of graphs uses rules to direct readers to scan the graphs sequentially by row.

Keep in mind that rules and grid lines should always be rendered subtly, never prominently in the form of thick, dark, or bright lines.

Other Arrangements of Multi-Graph Series

Graphs in a series of small multiples like those that we've already examined all share the same categorical and quantitative scales. The same unit of measure is used in each graph, such as sales in dollars, order count, expenses in dollars, or headcount. Two variations on the small multiples theme are also useful. When you have a matrix of graphs (rather than a single series of small multiples that vary along one variable only, such as department in *Figure 11.15*), the graphs can vary along two variables simultaneously if you assign one variable to the columns and the other to the rows. This arrangement can be thought of as a *visual crosstab* or *visual pivot table*. This is easier to explain by showing. Here's a simple example:

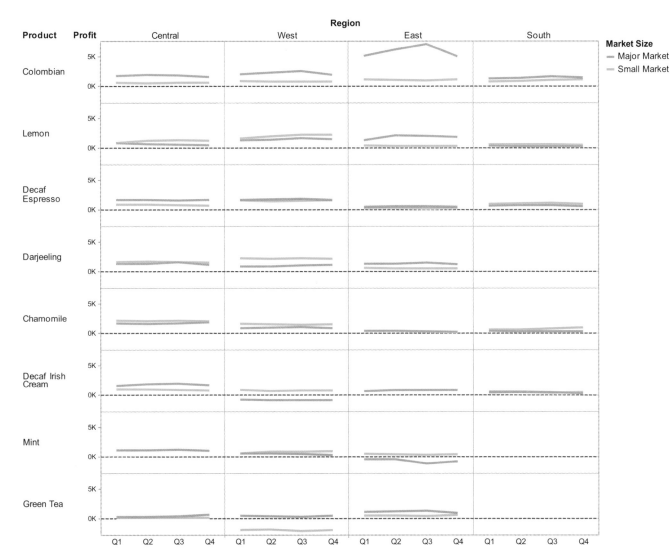

A normal series of small multiples introduces one additional categorical variable to the display; multiples arranged in a visual crosstab introduce two variables.

FIGURE 11.16 This matrix of graphs, created using Tableau, displays products by row and regions by column.

It is also useful at times to compare multiple units of measure. Imagine that you want to display forecast versus actual revenue, order count, and average order size across four quarters of a year for five separate products. You want to arrange these separate measures of sales in a way that enables easy comparisons. A matrix is an ideal solution. Take a look:

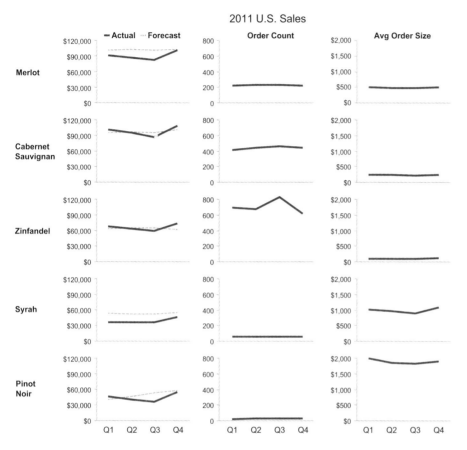

FIGURE 11.17 This matrix of graphs includes three separate but related series: forecast versus actual, order count, and average order size.

Note that a matrix of graphs consisting of multiple but related series must share a categorical variable, which in this case is product. Otherwise, the series of graphs wouldn't be related. This type of design allows readers to examine each series of sales measures independently (down a single column), examine all sales measures for a single product (across a single row), or view both at the same time looking for overall patterns. This is a powerful method for quantitative communication.

Summary at a Glance

Combining Multiple Units of Measure

When you wish to display two units of measure for purposes of comparison, the best way to avoid confusion is usually to use separate graphs rather than a single graph with two quantitative scales.

Combining Multiple Graphs in a Series

When you need to add one more variable (i.e., another set of categorical subdivisions) to a graph, but you've already used all the practical means to visually encode values in it, you can do so by constructing a series of related graphs, in which each graph in the series displays a different instance of the added variable.

Topic	Practices
Consistency	• Graphs in a series of small multiples should be consistently designed with only one exception: text used for labels, titles, or legends does not need to appear redundantly in each graph.
Arrangement	• Arrange the graphs in a series of small multiples in the way that makes it as easy as possible to focus on and compare the values that are most relevant to your readers' interests.
Sequence	• If the index variable has an intrinsic order, you should sequence the graphs in this order unless you wish to display a ranking relationship. • Otherwise, rank the graphs in order based on a quantitative measure associated with the index variable.

Rules and grid lines	• Only use rules or grid lines between graphs in a series when either of these two conditions exists:
	  ○ The graphs must be positioned so closely together that white space alone cannot adequately delineate them.
	  ○ The graphs are arranged in a matrix and are positioned so closely together that white space alone cannot adequately direct your readers to scan either across or down in the manner you intend.

12 SILLY GRAPHS THAT ARE BEST FORSAKEN

Several graphs that are readily available in software fail miserably at data presentation even though their popularity is growing. The stories that people attempt to tell with these graphs can be told simply and clearly using alternatives that are described in this chapter.

In this chapter we'll take a look at several graphs that are almost never useful. The reason is straightforward: they don't communicate data effectively. In every case the stories that these graphs attempt to tell can be told quite well using other graphs. We'll examine the following seven graphs—why they don't work and how their stories can be better told:

- Donut charts
- Radar charts
- Area charts for combining part-to-whole and time-series relationships
- Circle charts
- Unit charts
- Funnel charts
- Waterfall charts for simple part-to-whole relationships

These graphs are sometimes used because they're cute, sometimes because they're based on a meaningful metaphor (e.g., funnel charts), and often for the simple reason that they are readily available in software. No matter how convenient these graphs are or how impressive they seem—they might even be beautiful—they fail if they don't present information clearly, accurately, efficiently, and accessibly. They all fail for reasons, mostly perceptual, that we've considered in earlier chapters.

Donut Charts

A *donut chart* is a pie chart with a hole in the middle. Both pie and donut charts attempt to display parts of a whole. If pie charts are graphical pastries filled with empty calories, donut charts are the same and more. Here's a typical example:

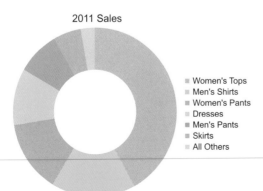

2011 Sales

- Women's Tops
- Men's Shirts
- Women's Pants
- Dresses
- Men's Pants
- Skirts
- All Others

FIGURE 12.1 This is a typical donut chart.

You might argue that the legend could be eliminated by labeling the sections of the donut directly. This would definitely improve the chart. We would still be left, however, with the difficult task of comparing the areas of the donut's sections or the distances along its inner or outer edges to make sense of this chart. To this you might respond that the value of each section could be displayed in it as text, thus relieving us of these difficult comparisons. To make this case, however, is to admit that the chart doesn't work graphically. If we need to display the values as text, why use a graph at all? Why not use a simple table instead? Or why not use a bar graph?

Just as the part-to-whole stories of pie charts can be better told with bar graphs, the same is true of donut charts. Here's the data from the donut chart displayed as a bar graph:

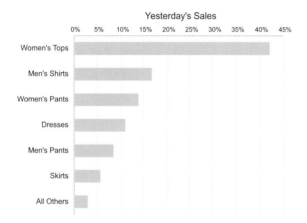

FIGURE 12.2 This bar graph displays the data from *Figure 12.1* in a way that can be easily read and compared.

The donut chart is definitely one that people are tempted to use because it's cute. Software vendors sometimes make donut charts enticing enough to eat (see below), but they'll still leave you empty.

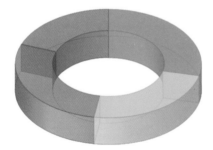

FIGURE 12.3 This example of an eye-catching but silly donut chart is offered by .net Charting.

Radar Charts

A *radar chart*, also called a *spider chart*, is laid out in a circular fashion, rather than the more common linear arrangement. It consists of axis lines that start in the center of a circle and extend to its periphery. Each axis can represent either an independent measure related to a single thing (for example, different measures of a cereal's nutritional content, such as protein, fat, sugar, potassium, and calories) or a single measure broken into multiple categorical items (such as expenses per department). The axes in the following example are of the latter type, each representing a different sales channel for a single measure: sales revenue.

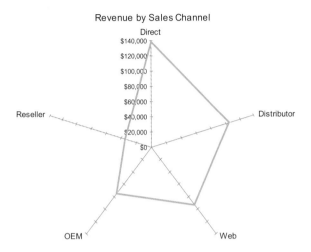

FIGURE 12.4 This is a typical radar chart.

The quantitative scales that run along the axes are generally arranged to begin in the center with the lowest value and extend toward the outside with increasing values. The lines that connect the individual values on each axis form a polygon, which is sometimes filled in with color. The data displayed in this example would ordinarily be presented in a bar graph, as shown below. Take a moment to compare the relative ease with which you can read these two graphs. Although the radar graph certainly looks interesting—much cooler than the more familiar bar graph—it takes us longer to compare the sales of the various sales channels in the radar graph. Positions along a quantitative scale are much easier to compare when they are laid out linearly along a single vertical or horizontal axis. Also, in a case like this when additional meaning can be displayed by ranking the items, a bar graph supports this nicely, but a radar graph does not because it isn't clear where the information begins and where it ends or whether it should be read clockwise or counterclockwise.

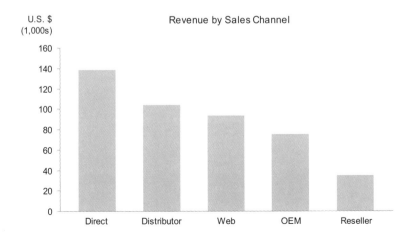

FIGURE 12.5 The bar graph displays the same information as *Figure 12.4*.

Radar charts are not limited to a single series of values. In the following example, sales are divided into three product types, one per line:

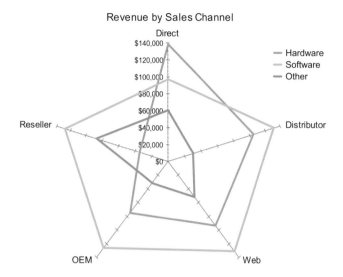

FIGURE 12.6 This radar chart contains three sets of value.

Similar to other types of graphs, a radar chart becomes more difficult to read as more data sets are added, and the problem is magnified by its circular design, which slows and complicates scans and comparisons.

Displays involving time are the one common instance in which the circular nature of a radar chart actually lends itself to the nature of the data that it displays. Because we perceive time as cyclical—a series of years, quarters, month, days, or hours that repeat over and over—a circle may be used to represent it. We are all accustomed to the hands of a clock turning in circles to display minutes of the hour and hours of the day, so a chart like the one below fits this familiar arrangement.

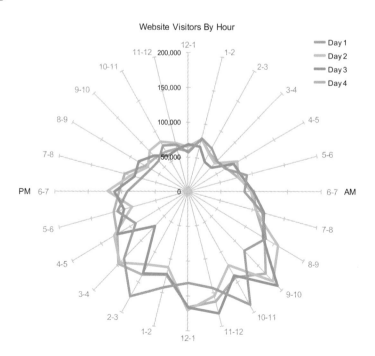

FIGURE 12.7 This radar chart displays change through time.

Although this works, most of us find a linear version, as shown below in the form of a line graph, easier to read and interpret:

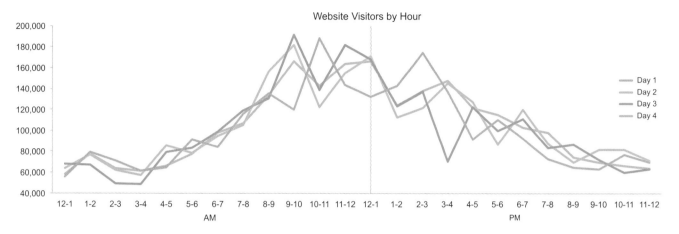

For this reason, make sure a circular graph offers clear advantages before you use it in place of a simple, clear, and familiar line graph.

FIGURE 12.8 The same time series that appears in *Figure 12.7* is displayed here as a line graph.

Stacked Area Graphs for Combining Part-to-Whole and Time-Series Relationships

In general, any graph that encodes values using the 2-D areas of objects, including pie charts, is an *area graph*. We avoid using area graphs whenever possible because visual perception in humans can only compare areas as rough estimates. In this section we'll look at a specific type of area graph that is often used to show how the total of something and its parts change through time. Here's a simple example:

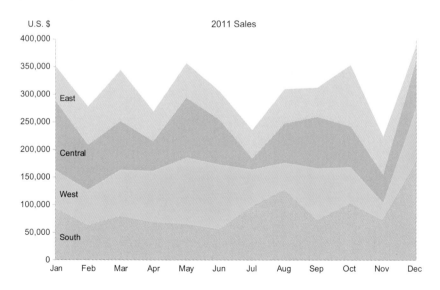

FIGURE 12.9 Stacked area graphs such as this are often used to display how a total and its parts change through time.

Because stacked area graphs are like line graphs in most respects, we tend to read them as we do line graphs; this tendency gets us into trouble. We can follow the pattern formed along the top of the chart (the top of the pink section) to examine total monthly sales, and we can follow the pattern formed along the top of the blue section to trace monthly sales in the South, but we cannot follow the tops of the orange, green, or pink sections to follow monthly sales in the West, Central, or East regions. This is because the slopes of the lines at the tops of these areas are influenced by the slopes of the areas below them and therefore don't represent the patterns of change for those regions. To see the pattern of change for the West region, we must look only at the vertical distance between the bottom and top of the orange section above each month label, and try to imagine those values as a line going up and down. Here's what the pattern of monthly sales for the West looks like when shown by itself:

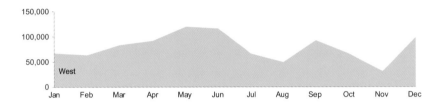

FIGURE 12.10 This graph displays sales in the West region only, taken from *Figure 12.9*.

Constructing this pattern in your head by viewing the West region in *Figure 12.9* simply isn't possible unless you're an extremely rare savant. Because of this difficulty, stacked area graphs suffice only when you most want to show how the whole changes through time while giving your readers a rough sense of how the parts compare to one another in their contributions to the whole. Even on such occasions, make sure that your readers know not to follow the slopes of the lines along the tops of each section of color to see how a part of the whole changed through time.

The best way to show how a total and its parts change through time usually involves two graphs: one for the whole and one for its parts. Here's a dual-graph solution for the same sales data as before:

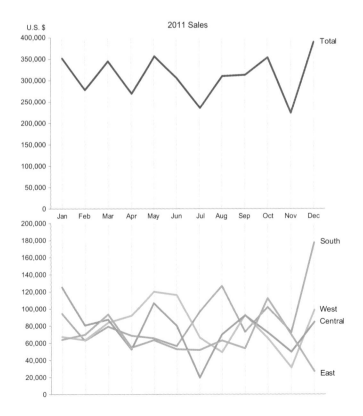

FIGURE 12.11 These graphs collaborate to display the sales information that appears in *Figure 12.9*.

Now, we can clearly see how the whole and its parts change through time, individually and in relation to one another. As I've said before, quantitative stories cannot and need not always be told with a single graph. Clarity sometimes demands a little more space.

Circle Charts

We sometimes use circles to encode data in graphs, even though they require 2-D area comparisons, because better methods such as 2-D position have already been used. For instance, we use circles on maps and in bubble charts. To use them when a perceptually more effective means is available doesn't make sense if we wish to communicate clearly, however. Here's an example that I found about the financial meltdown that occurred in 2007:

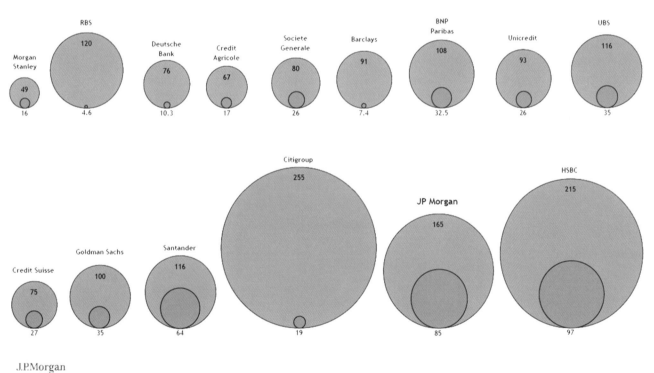

Why would J.P. Morgan tell the story of the financial meltdown's impact on banks in this way? People sometimes do this because they've seen similar charts in newspapers or on the web and are naively following a bad example. People who know better sometimes do this because they care more about visual impact than clarity of communication. If we want to tell the truth in a way that people can easily understand, this is not an effective approach.

The point of the story that J. P. Morgan was trying to tell was that it lost relatively little compared to other banks. When the loss is expressed as a

FIGURE 12.12 This chart uses circles to encode pre- and post-meltdown market values of 15 banks.

percentage of remaining value on January 20, 2009 compared to value in Quarter 2 of 2007, then only Santander suffered a smaller loss than J.P Morgan. Nothing about the design of this chart features this message, however. The use of circles—blue for before the meltdown and green for after—makes even approximate comparisons difficult, so we're forced to read the numbers. It's actually worse than it seems on the surface because the person who created this chart didn't size the area of the circles to represent the differences in values but mistakenly used the diameters instead, which underrepresents the differences dramatically.

This is a shame because this particular story is incredibly easy to tell graphically. These before-and-after values can be clearly shown and compared using bars. Here's one of several possible designs:

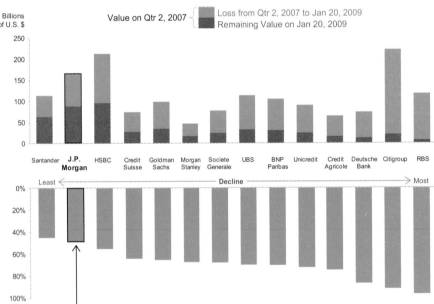

FIGURE 12.13 Redesign of *Figure 12.12* as two bar graphs.

Now, the fact that the story is featuring J.P. Morgan is apparent, pre- versus post-meltdown values can be easily compared in the upper graph, and the fact that J.P. Morgan suffered relatively little decline in value, second only to Santander, is obvious. The dual-graph solution is simple to construct following the guidelines that we've covered in this book.

Unit Charts

Just as pie charts work well for teaching children the concept of fractions but not for displaying parts of a whole in a discernible manner, *unit charts* might be useful for teaching children to count, but, for most people over six years of age, they're not particularly effective. What is a unit chart? The term might not be familiar, but you've probably seen them many times. Here's a typical example:

STUDENT SELF-REPORTED DRINKING

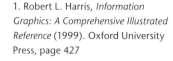

Never tried alcohol 6%

Tried alcohol,
currently don't drink 13.3%

Light drinker 29.1%

Moderate drinker 44.7%

Heavy drinker 6.9%

FIGURE 12.14 From a larger infographic titled "#1 Party School," created by E. J. Fox.

According to Robert L. Harris, a unit chart is defined as follows:

A chart used to communicate quantities of things by making the number of symbols on the chart proportional to the quantity of things being represented. For example, if one symbol represents ten cars and five symbols are shown, the viewer mentally multiples ten times five and concludes that the group of symbols represented 50 actual cars. Simple geometric shapes or irregular shapes such as pictures and icons are generally used. Each provides basically the same degree of accuracy. When the symbols are geometric shapes, the chart is occasionally called a black chart. When pictures, sketches, or icons are used, the chart is often referred to as a pictorial unit chart. Unit charts are used almost exclusively in presentations and publications such as newspapers, magazines, and advertisements.[1]

1. Robert L. Harris, *Information Graphics: A Comprehensive Illustrated Reference* (1999). Oxford University Press, page 427

In the example above, a 10x10 matrix of dots with one missing for a total of 99 was used to display self-reported student drinking habits. Each dot represents 1% out of 100% of students. This chart would give a 1st grader who's interested in student drinking habits an opportunity to understand the data while practicing his counting skills. Of course, relatively few 1st graders are going to find this information useful. This information is only of interest to adults—people who don't usually need to practice counting. Here's the same information displayed as a bar graph.

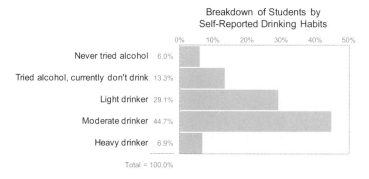

Breakdown of Students by
Self-Reported Drinking Habits

FIGURE 12.15 A redesign of *Figure 12.14* as a bar graph.

Now, rather than counting, we can easily use our perception of length differences to compare the bars with speed and precision. Why would anyone ever use unit charts to display quantitative data? For some reason journalists seem to love them. They find the conceptual simplicity of unit charts appealing, perhaps because they have a low opinion of their readers' intelligence.

The simplest form of a unit chart displays a single row or column of units for each item rather than a matrix of both as we saw in the previous example. As you can see in the example below, a unit chart of this type is easier to read:

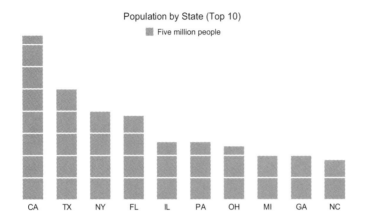

FIGURE 12.16 This is an example of a simpler unit chart.

Are one-dimensional unit charts such as the above worthwhile? We can read them much as we read bar graphs, with one minor difference—the segmentation of values into units inclines us to slow down and count, as opposed to the simpler, faster task of comparing heights of the bars in relation to a quantitative scale, which is missing. Not a big problem, some might argue, but significant enough to discourage their use when better means are available. The example above illustrates the common practice of including partial units—in this case rectangles with less height than a full unit—to represent partial quantities. For example, in this chart, state populations have been rounded to the nearest million people, reducing the heights of those uppermost rectangles that represent less than five million people.

As Harris pointed out in his definition, unit charts come in two basic types: those that use geometric shapes such as rectangles or circles, and those that use irregular shapes such as pictures or icons. Here's an example of the latter:

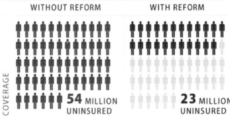

FIGURE 12.17 This unit chart uses icons to represent the values.

The simple icon that represents one or more people (in this case one million) is a staple of journalistic unit charts. Although this familiar icon instantly clues readers into the fact that people are the subject of the display, the values cannot be discerned without counting the icons or more simply reading the numbers (for example, 54 million uninsured). Because we cannot preattentively compare counts that exceed three or four objects, we're forced to abandon rapid visual

perception and rely on slower methods of discernment—counting or reading—
to understand this graph.

Unit charts are often complicated by the existence of irregular numbers of
columns, rows, or items within them. The next example from a larger infographic depicting volcanic eruptions illustrates this problem:

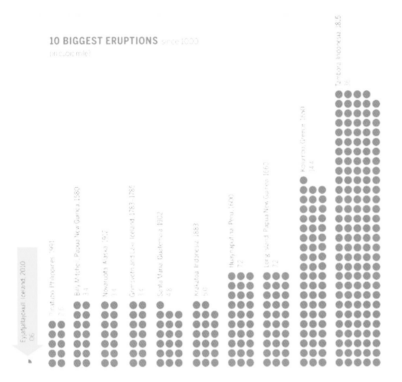

FIGURE 12.18 Created by Steven
Dutch, Natural and Applied
Sciences, University of Wisconsin –
Green Bay.

The columns of units range from one for Iceland on the left to five for Indonesia
on the right, and the number of units per row sometimes fill the columns
equally (e.g., Papua New Guinea) and sometimes fill unequal portions of them
(e.g., Greece). This makes it impossible to compare the size of eruptions to get
even a rough sense of the differences by comparing the heights of the separate
series. This forces us to either count, read the numbers, or do our best to compare the areas formed by each—an ill-fated venture.

The sense of simplicity that unit charts convey conflicts with the complication that they impose on perception. As such, they are a cute but poor form of
communication.

Funnel Charts

When businesses monitor sales, they often track their progress on a daily basis,
watching sales opportunities work their way through a series of stages that's
called the sales funnel because sales decrease with each stage, like the shape of a
funnel, until the final stage when actual purchases are made by a subset of the
initial sales opportunities. Similar to the way that the car dashboard metaphor
has inclined people to populate information dashboards with silly gauges and
meters, the funnel metaphor has given rise to a silly array of funnel-shaped

quantitative displays. Despite their primary use for sales tracking, they can be used for any series of quantitative values that are whittled down by stages in a process. They are also frequently misused as an alternative to pie charts for displaying parts of a whole. The following example tracks stages in the process of website visitors making a series of choices to become increasingly involved in the site:

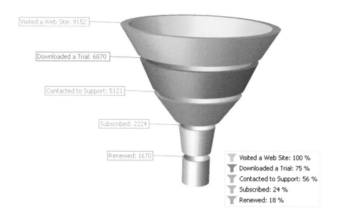

FIGURE 12.19 An example of a funnel chart found at www.devexpress.com.

The primary problem with funnel charts is that they impose the shape of an actual funnel in a way that doesn't correspond to the values. In rare cases when the sizes of each funnel section actually does correspond to the values, the funnel ends up appearing oddly unfunnel-like in shape, and the accurate sizes don't really help anyway because we can't compare 2-D areas effectively even when they're shaped alike, such as slices in a pie chart.

There is no question that it's useful to track a diminishing series of values in a graphical form. The funnel metaphor is a good one, but the actual image of a funnel cannot be used to represent the values effectively. Why not stick with the metaphor but use a normal bar graph to display the values, as shown below?

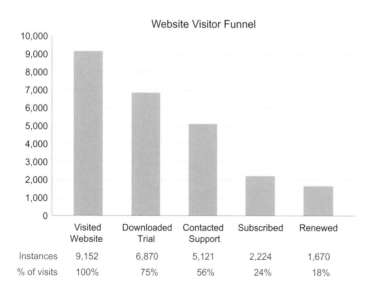

FIGURE 12.20 This bar graph displays the same funnel data as *Figure 12.19*.

I arranged the values horizontally from left to right because that is a more intuitive representation of events that occur in time. Now the data-encoding objects—bars in this case rather than sections of a funnel—show the values in a way that can easily and efficiently be compared. Anyone who cares about the actual values in a funnel process will appreciate the effectiveness of this form of display.

Waterfall Charts for Simple Part-to-Whole Relationships

Waterfall charts can be quite effective in telling a particular story that involves how increases and decreases in a balance affect that balance through time. I constructed the following example to illustrate an occasion when a waterfall chart provides a good solution:

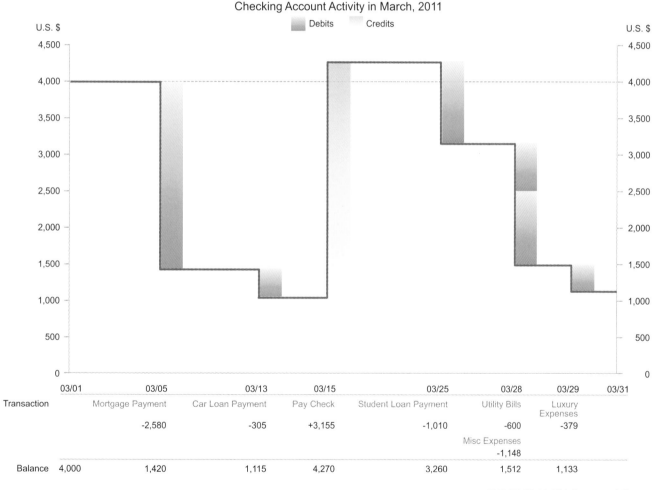

I had to construct this example manually because I'm not aware of any software that can produce a waterfall chart such as this. The chart tells the story of how a checking account's balance changed at particular points in time as a result of credit and debit transactions. The information is presented in a way that pinpoints the dates of transactions. Transactions can be easily spotted, identified

FIGURE 12.21 This is a waterfall chart that tells its story effectively.

as credits or debits, and compared. The balance is featured in a way that is easy to track through time. The transactions have been described precisely in tabular entries below the chart. This example thoroughly leverages the strengths of a waterfall chart.

Unfortunately, few of the waterfall charts that I've seen are used to tell this particular type of story, of a changing balance through time. Most are little more than a vertical stacked bar with its segments staggered in space horizontally, such as the following example:

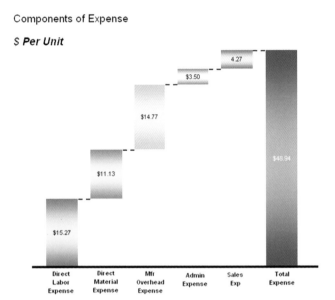

FIGURE 12.22 This example from www.appiananalytics.com is typical of the way waterfall charts are used.

This is nothing more than a stacked bar (the five segments combine to form the final bar on the right) that has been exploded horizontally. The sequence from left to right does not represent change through time. Stacked bars are only useful when several of them are used in a graph to feature the totals of several items (e.g., the expenses of several departments) subdivided into parts (e.g., different expense categories) for the primary purpose of comparing the totals and the secondary purpose of providing a rough sense of the parts. Just as with most part-to-whole stories, the one on the pseudo-waterfall chart above can be told more effectively using a regular bar graph, such as the one below:

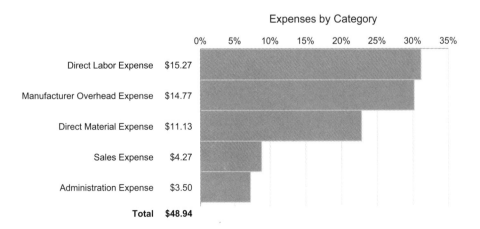

FIGURE 12.23 This bar graph shows the same part-to-whole information as the pseudo-waterfall chart in *Figure 12.22.*

Other silly graphs are out there as well, along with silly uses of otherwise effective graphs. We've only examined the ones that I encounter often. Just because a graph of a particular type exists in your software doesn't mean it's useful. It would be nice if we could trust software vendors to provide only what's useful, but most of them will provide anything they think will sell, useful or not. Good graph choices are up to us.

PRACTICE IN GRAPH DESIGN

You've come far in your expedition into the world of graph design. It's now time for some practice to pull together and reinforce all that you've learned. Expert graph design requires that you adapt and apply what you've learned to a variety of real-world communication problems. Working through a few scenarios with a clear focus on the principles of effective graph design will strengthen your expertise and your confidence as well.

Exercise #1

I found a graph like the one pictured below in the user documentation of a popular software product that specializes in data analysis and reporting. It displays quarterly revenue and the number of guests at a particular hotel. The graph appeared as an example of the proper use of scatter plots. I made no attempt while recreating this graph to make it any worse than it already was. Spend some time examining it, identify one or more stories that it tells, and then follow the instructions below.

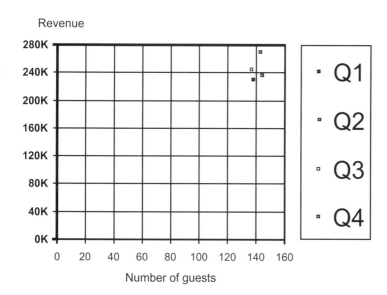

List each of the problems that you detect in the design of this graph:

Now, suggest a solution to each of these problems:

Exercise #2

This example was derived from the user documentation of the same product that gave us the graph for Exercise #1. (I've found that I can't invent examples that are as poorly designed as those that I find quite readily in actual use.) As you can see, the intention here is to display sales revenue in the state of Kansas associated with 12 products across the four quarters of a single year. Once again, look at the graph closely and follow the instructions below.

Kansas

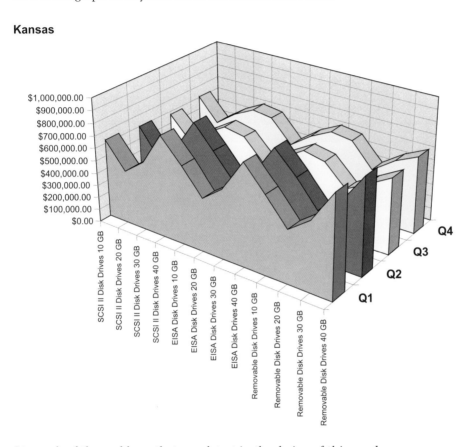

List each of the problems that you detect in the design of this graph:

Now, suggest a solution to each of these problems:

Exercise #3

Design a better way to display the data below, taken from the November 29, 2004 issue of *CRN* magazine. Read the paragraph to the left of the graphic to learn the story that the graphic is supposed to tell, and then answer the questions below.

List each of the problems that you detect in the design of this graph:

Now, suggest a solution to each of these problems:

Exercise #4

The primary intention of this graph is to display how the average selling price of gizmos changed from month to month throughout the year; the secondary intention is to relate the average selling price to the range of prices during those same months. Given these objectives, examine the graph, and respond to the instructions below.

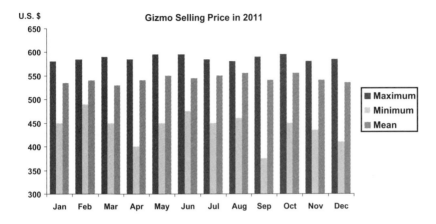

List each of the problems that you detect in the design of this graph:

Now, suggest a solution to each of these problems:

Exercise #5

It's now time to redirect your attention closer to home. Select three graphs that are used at your place of work. Make sure that at least one of them is a graph that you created. For each of the graphs, respond to the following instructions:

Graph #1

List each of the problems that you detect in the design of this graph:

Now, suggest a solution to each of these problems:

Graph #2

List each of the problems that you detect in the design of this graph:

Now, suggest a solution to each of these problems:

Graph #3

List each of the problems that you detect in the design of this graph:

Now, suggest a solution to each of these problems:

The remaining exercises ask you to design graphs from scratch to achieve a specific set of communication objectives. You may construct the graphs using any relevant software that is available to you, such as spreadsheet software.

Exercise #6

Imagine that you work in the Financial Planning and Analysis department. It's your job to report how actual quarterly expenses during the previous year compared to the budget for each of the five departments that report to the Vice President (VP) of Operations. The VP is only interested in the degree to which actual expenses deviated from budget, not the actual dollar amount. Here's the raw information:

Department	Q1 Budget	Q1 Actual	Q2 Budget	Q2 Actual	Q3 Budget	Q3 Actual	Q4 Budget	Q4 Actual
Distribution	390,000	375,000	395,000	382,000	400,000	390,000	410,000	408,000
Facilities	675,000	693,000	800,000	837,000	750,000	713,000	750,000	790,000
Human Resources	350,000	346,000	350,000	342,000	350,000	340,000	350,000	367,000
Information Systems	950,000	925,000	850,000	890,000	875,000	976,000	900,000	930,000

Create one or more graphs to display this information. Once you've completed your graph, take a few minutes to describe its design, including your rationale for each design feature, in the space below.

Exercise #7

You are the company's marketing analyst. You have discovered through an analysis of last year's revenues that sales have significantly benefited from the use of television ads. You need to present this benefit to the Chief Executive Officer (CEO). Here's the raw information:

	Jan	Feb	Mar	Apr	May	Jun	Jul	Aug	Sep	Oct	Nov	Dec
TV Spots	20	25	20	30	30	30	30	20	15	25	30	35
Revenue	50,000	55,000	52,000	57,000	58,000	59,000	57,000	50,000	45,000	55,000	63,000	78,000

Create one or more graphs to display this information. Once you've completed your graph, take a few minutes to describe its design, including your rationale for each design feature, in the space below.

Exercise #8

For this last exercise, you are a Sales Analyst. Based on analysis of sales revenues for the past four years, you've noticed that there is a clear cyclical trend. The highest revenues are always generated during the last month of each quarter and the last quarter of each year, without exception. You want to use this information to point out the need for distributing sales activity more evenly across the year so that sales operations are not hit with such spikes in activity at quarter ends, especially at year ends. The following table contains the data, in U.S. dollars:

Year	Jan	Feb	Mar	Apr	May	Jun	Jul	Aug	Sep	Oct	Nov	Dec
2008	50,000	52,000	55,000	51,000	52,000	55,000	48,000	49,000	50,000	53,000	56,000	62,000
2009	54,000	56,000	57,000	53,000	53,000	57,000	53,000	55,000	59,000	60,000	65,000	71,000
2010	61,000	60,000	65,000	62,000	64,000	68,000	59,000	60,000	65,000	67,000	70,000	75,000
2011	60,000	61,000	68,000	63,000	63,000	67,000	60,000	61,000	67,000	66,000	68,000	74,000

Create one or more graphs to display this information. Once you've completed your graph, take a few minutes to describe its design, including your rationale for each design feature.

You can find answers to these exercises in Appendix H, *Answers to Practice in Graph Design*.

13 TELLING COMPELLING STORIES WITH NUMBERS

Important stories live in the numbers that measure what's going on in the world. Before we can present quantitative information, we must first uncover and understand its stories. Once we know the stories, we can tell them in ways that help others to understand them as well.

We are natural-born storytellers. Humans have been creating and preserving culture through stories since we learned to speak. The numbers that are stored in our databases and spreadsheets have important stories to tell, and only we can give them a voice. However, before stories can be told, they must be discovered and understood. Data sensemaking precedes data presentation. Before you present data to others, you should always ask: "What's the story?" The nature of the story and the nature of the audience will determine the best way to tell it. Once you know the story, presenting it is much like storytelling of all types; it boils down to clear communication. Telling stories with numbers—statistical narrative—involves a few specialized skills, but they're not complicated. In this chapter, we'll examine a few of the principles and practices that will help to bring quantitative stories to life.

In 2006, when Hans Rosling of www.GapMinder.org spoke at the TED Conference, it was probably the first time in history that an audience arose to its feet in applause after viewing a bubble plot.

For instruction in the use of graphs for data exploration and sensemaking, see my book *Now You See It: Simple Visualization Techniques for Data Analysis* (2009).

FIGURE 13.1 Hans Rosling of www.GapMinder.org, presenting for the first time at the TED Conference in 2006.

Rosling's bubble plot went further than most by displaying a fourth variable in a way that was novel: through animation. Bubbles moved around in the plot area to show how values changed through time. In this particular presentation, the bubble plot displayed a bubble for each country in the world, sized according to population and colored to group countries by region. Horizontal position along the X axis represented the average number of births per woman of childbearing age, and vertical position along the Y axis represented average life expectancy at birth in years. Beginning with United Nations data from 1962, the graph shows that countries at that time were roughly divided into two groups: developed countries in the upper left section of the graph, with few children per woman and long lives, versus developing countries in the lower right, with many children per woman and short lives. When the bubbles began to move around to show how these values changed from 1962 through 2003, Rosling drew the audience's attention to particular bubbles that exhibited interesting behavior, such as Bangladesh, which began a rapid decrease in the number of children per woman and increase in longevity when the imams in that country began to promote family planning. By the time the bubbles ceased their movement in the year 2003, the graph displayed a completely different world—one in which most countries had small families and long lives, with exceptions almost entirely among African countries where the HIV epidemic kept average lifespans short.

Why did people find Rosling's graph more engaging than most? I suspect for three reasons:

1. The story that he told was important; it revealed facts about life expectancy and family size that the audience could easily care about.
2. The story was told in a way that the audience could understand.
3. Rosling is a skilled storyteller.

Just as Rosling brings important quantitative stories to life in ways that enlighten and move his audience, you can find compelling ways to tell your own stories. If a story is worth telling, it's worth telling it well. To do so, you need a few skills that apply to communication in general, and a few that are specific to stories contained in numbers.

Learning these skills has, in recent years, become the interest of diverse groups that work with data, especially journalists. Journalistic infographics have attracted a great deal of attention recently, at times with breathtakingly informative visualizations but more often with silly eye-catching displays that dazzle while saying little and saying it poorly. Even Rosling was lured into this trap when he allowed the video crew of BBC4 to add video effects to one of his presentations that made it almost impossible to see and follow the data. The presentation was flashy, but the story got lost behind the glaring lights and visual effects.

FIGURE 13.2 Hans Rosling presenting information while standing behind a virtual graph placed on the screen by means of a video effects in a BBC4 documentary entitled "The Joy of Stats."

Had the video effects been eliminated, Rosling's presentation would have engaged viewers in the data, drawing them into the story, rather than distracting them with the special effects. Stories can be told in ways that thrill and move an audience without using silly tricks that obscure the information.

Characteristics of Well-Told Statistical Stories

Stories can be told in various ways involving multiple media. Words can be complemented by pictures and diagrams. You can rarely tell quantitative stories by expressing numbers as text alone; you must use pictures as well. Whether expressed as text or graphically, numbers are numbers, but text and graphics each possess different strengths. Visual representations of numbers feature patterns, trends, and exceptions and make it possible for people to compare whole series of values to one another, such as domestic versus international sales throughout the year. You must learn when words work best and when graphs and other forms of visual display are needed, and then interweave them in powerful, engaging, and informative ways.

Let's examine several of the principles and practices that are required to tell stories with numbers in ways that inform and matter. Statistical narrative usually works best when it exhibits the following characteristics:

- Simple
- Seamless
- Informative
- True
- Contextual
- Familiar
- Concrete
- Personal
- Emotional
- Actionable
- Sequential

Simple

A story is more likely to take hold in the minds of your audience if you tell it simply. You must identify the essential message and then present it as simply as possible without distraction. Simplicity isn't always easy. In fact, it is often quite difficult to achieve. You must carefully discern the vital beating heart of your message—its essence—before constructing your story and crafting its presentation. Once your vision is clear, you must then find the clearest possible way to present it, both in content and means of expression.

Paul Grice, a philosopher of language, developed a series of "conversational maxims"—rules for communicating to one another courteously and effectively. Among his guidelines are these two:

- Make your contribution as informative as is required for the current purposes of the exchange.
- Do not make your contribution more informative than is required.

When telling a story, you should present all that must be known to get the message across, but no more. The effectiveness of the story relies on a fine balance between too little and too much. Comic book artist Scott McCloud explains how he crafts his stories to achieve "amplification through simplification"—amplification of meaning and understanding by simplifying the means of expression. He illustrates the concept in the following panel:

FIGURE 13.3 An excerpt from Scott McCloud (1994) *Understanding Comics: The Invisible Art.* HarperCollins, page 30.

Simplification involves a stripping away of all that's non-essential, which allows your audience to easily focus without distraction on what's most important. As Leonardo da Vinci wisely recognized, "Simplicity is the ultimate sophistication."

Seamless

Unlike some stories, statistical narrative almost always involves integrating words and images. The patterns, trends, and exceptions that constitute the meanings in data must be given visual form, for they are difficult—sometimes impossible—to weave together in our heads from words and tables of numbers. Verbal and visual forms of expression are both languages of sorts. Just as letters combine to form words, and words combine to form sentences, graphical marks combine to form visual objects, and those objects combine to form more complex objects such as graphs, which we can use to tell quantitative stories. As Jacques Bertin taught in his groundbreaking book *Semiologie graphique* (*Semiology of Graphics*) and Robert Horn extended in *Visual Language: Global Communication for the 21st Century*, graphics are a specialized language that can communicate quite effectively if we learn its rules of syntax and semantics.

Edward Tufte has long taught the power of seamlessly integrating words and pictures to tell stories. He and I both labor hard to interweave graphs and other images with the text that describes them, rather than the easier (and cheaper) method of referencing figures by number and placing them wherever they most conveniently fit. Verbal language expresses some information best, but other information is better communicated visually. Words and images collaborate elegantly when we resist arbitrarily separating them. Tufte says:

> *Data graphics are paragraphs about data and should be treated as such.*

> *Words, graphics, and tables are different mechanisms with but a single purpose—the presentation of information. Why should the flow of information be broken up into different places on the page because the information is packaged one way or another?*[1]

One of the finest early examples of this seamless tapestry of words and pictures comes from the writings of Leonardo da Vinci:

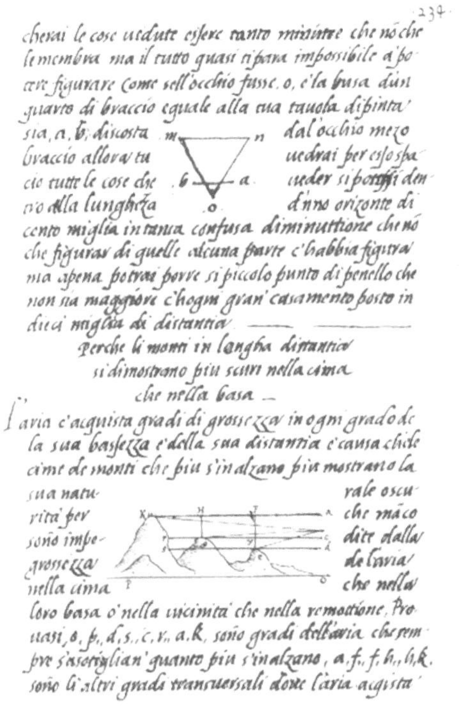

FIGURE 13.4 A seamless integration of words and pictures from Leonardo da Vinci's notebooks, later published under the title *A Treatise on Painting*.

1. Edward Tufte (2001) *The Visual Display of Quantitative Information*, Second Edition. Graphics Press, page 181.

As you can see, da Vinci saw no reason to arbitrarily separate words and pictures that are meant to work together.

When you combine words and pictures to tell a story to a live audience or by means of an audio-video recording, you can speak the words and show the pictures simultaneously. If the images are projected on a screen, you can gesture to those parts that you want your audience to focus on in any given moment. If you speak the words, there is no reason to display them as text on the screen as well. In fact, doing so interrupts the flow of communication. When your audience attempts to both read words on the screen and listen to the words coming from your mouth, the information gets into their brains less effectively than words coming through one channel only. Don't force your audience to read when they could be using their eyes to contemplate an image that complements your words. Visual and verbal content are processed by different parts of the brain and can be perceived and processed together with ease, resulting in richer understanding.

Informative

Stories should inform, revealing facts or interpretations of facts that your audience doesn't already know. Few experiences are more sleep-inducing than sitting through a presentation that repeats what you already know. If you want to arouse your audience's curiosity and interest, give their brains something new to chew on.

In their marvelous book *Made to Stick*,[2] the brothers Chip and Dan Heath describe what's needed to get a message to stick: to communicate it in a way that is understood, remembered, and has a chance to produce the hoped-for response. Several of the points that I'm making in this chapter are eloquently described in *Made to Stick*. One way to make a message stick is to elicit surprise by revealing something unexpected. This can often be done with images that force people to look at something from a new perspective. Hans Rosling's 2006 presentation at the TED Conference stuck, in part because he revealed a relationship between fertility (measured as the average number of children per woman of child bearing age) and life expectancy at birth (measured as the average number of years a child will live) that surprised many in the audience. Afterwards, they left the auditorium informed and thoughtful.

2. Chip Heath and Dan Heath (2007) *Made to Stick: Why Some Ideas Survive and Others Die*. Random House.

True

Let's go no further without acknowledging the importance of truth. Stories need not be true to stick—history is strewn with the destructive fallout of lies—but they should be true if you hope to make the world a better place. If your audience perceives the story as true, they will care about it more and are more likely to respond. This means that the story should not only be true, but should also be perceived as true. You should validate the story with relevant evidence. You should name your sources to lend credibility both through transparency and by virtue of the source's perceived credibility. If you yourself believe the story to be true, convinced through solid evidence, your conviction will be noticed. If you have integrity and consistently exhibit it through your actions, the trust that you earn will extend to your stories as well.

Contextual

Quantitative stories cannot be told effectively merely by presenting numbers. Numbers alone—even those that measure something worthwhile—are meaningless unless you present them in context. In part, this means that you should reveal the pedigree of the numbers (that is, where they came from and how you might have adjusted them). Even more importantly, however, this means that you must provide additional information to which the numbers can be compared. People discover meanings in numbers primarily by comparing them to other numbers, and by comparing the patterns, trends, and exceptions that live within them to those that dwell in others. Relevant comparisons make numbers meaningful in a way that can be used to form judgments, make decisions, and take action.

Appropriate comparisons span a long list of possibilities. One of the most meaningful and informative involves comparing what's going on today to what happened in the past. History provides context that we dare not ignore, lest we relive its dark moments. Other examples of useful objects of comparison include:

- Targets (e.g., budgets and plans)
- Forecasts (i.e., what you predict will happen)
- Other things in the same category (e.g., comparing a product to other products, or comparing your company to competitors)
- Norms (e.g., comparing the performance of one hospital to the average performance of all hospitals in the cohort group)

Is it enough to tell the Vice President of Sales that $8,302,563.38 in revenue has come in so far this year? What does that mean? Let's add some context:

FIGURE 13.5 Several months of history has been added to provide context for current revenues.

Now, with monthly revenue values so far this year and only one more day left in July, it is easy to see that revenues have leveled off in this month after two months of significant increase. Let's add some more context:

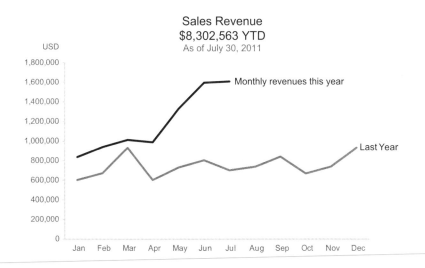

FIGURE 13.6 Last year's revenues have now been added to provide even more historical context.

With the addition of last year's revenues, it is now easy to see that the pattern of sales this year is quite different from the pattern last year. The fact that revenues leveled off this year in July doesn't look so bad in light of the fact that they decreased in July last year. Let's add one final bit of context:

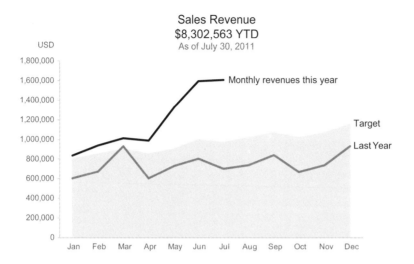

FIGURE 13.7 Target revenues have now been added to provide richer context for understanding revenue performance.

For quantitative information, meaning only becomes visible against the backdrop of comparisons.

Familiar

Know your audience. It's no secret that you must have a sense of the people you're addressing and customize the presentation to reach them. I don't speak to trained statisticians in exactly the same way that I speak to non-statisticians. I don't speak to doctors and nurses the same as I speak to people who work in a marketing department. I might be telling the same essential story to both groups, but I'll choose some of my words and images differently. Why? I want to express the story in familiar and relevant terms so that my audience will understand and care about it.

If you spend your life in a cloister without interaction with those outside the walls, you'll struggle to talk to those who live in the nearby village. If you're an expert data analyst, you probably speak and think in ways that aren't familiar to others in your organization who lack your training. If you can tell your story using simple words and images that are familiar to everyone, why would you use cryptic terms and complicated graphs, unless you're trying to impress them with your superior knowledge? If people care about the story, they will be much more impressed if you tell it in a way that they understand. Although Hans Rosling is a university professor who could confound most of us with the specialized knowledge of his field, we respect his work because he cares enough to boil his stories down (without watering them down) to simple terms. People who can't tell their stories in understandable ways are either naive (unaware of the world outside of their own small spheres), lazy (unwilling to craft the story in familiar terms), full of themselves (more interested in impressing than communicating), unskilled in the use of everyday language, or just don't understand their stories well enough to tell them clearly.

Concrete

Even though humans have evolved the extraordinary ability to think abstractly, we don't dwell entirely in the abstract. When our stories involve abstract concepts, which they often do, we can help people make sense of those abstractions by using concrete examples and metaphors. Chip and Dan Heath say it well:

> *Concrete language helps people, especially novices, understand new concepts. Abstraction is the luxury of the expert. If you've got to teach an idea to a room full of people, and you aren't certain what they know, concreteness is the only safe language.*[3]

Many organizations express their mission statements in shallow, ethereal, and uninspiring abstractions. Smart organizations express them in concrete terms that people can embrace with their heads and hearts. Some people intentionally use abstractions to hide the truth. Read the following excerpt from Enron's annual report for the year 2000:

> *Enron's performance in 2000 was a success by any measure, as we continued to outdistance the competition and solidify our leadership in each of our major businesses. We have robust networks of strategic assets that we own or have contractual access to, which give us greater flexibility and speed to reliably deliver widespread logistical solutions...We have metamorphosed from an asset-based pipeline and power generating company to a marketing and logistics company whose biggest assets are its well-established business approach and its innovative people.*[4]

We now know that, had the story been told in concrete ways in that annual report, employees could have willingly left in disgust before the company's implosion. The leaders of Enron intentionally kept the story abstract. They weren't trying to communicate; they were trying to obfuscate.

Some numbers are difficult to understand. This is especially true of extremely large and extremely small numbers. You can make it possible for people to wrap their heads around such numbers by expressing them in concrete terms, usually by comparing them to other numbers that are familiar and easy to understand. In the year 2010, the richest person in the world was Carlos Slim Helu, a Mexican Telecommunications mogul, whose net worth came to a total of $53.5 billion (USD). Most of us can't think in terms of this much money. Expressed in this way, the number is almost meaningless. One way to express the amount more concretely is to compare it to numbers that are more comprehensible, such as:

> Carlos Slim Helu's net worth of $53.5 billion (USD) is equal to the annual income of 1.7 million workers in the U.S. or 9.2 million workers in Mexico.

Even though we're still using numbers in the millions, they're within reach and concrete.

3. Chip Heath and Dan Heath (2007) *Made to Stick*. Random House, page 104.

4. Brian Fugere, Chelsea Hardaway & Jon Warshawsky (2005) *Why Business People Speak Like Idiots*. Free Press, page 11.

Personal

Stories have a greater effect when people connect with them in personal ways. Sometimes it's easy to make stories personal because they relate to your audience directly. In such cases, you can point out in concrete ways how the story addresses their interests. Even when the story is about people and things taking place in some remote part of the planet, there is usually a way to help people connect in some personal way if the story's important. The fact that global warming will literally bury beneath the sea islands where people now live might not affect us directly, but if you show where those islands are on a map along with pictures of those people, your audience might begin to care about those strangers and imagine facing their plight.

Emotional

The emotional element of an effective story is intimately related to the personal characteristic we just discussed. When things are personal, they stir up emotions in us; when our emotions are stirred, we tend to experience things personally. As much as we'd like to believe that we're rational creatures, research has shown that most decisions are based on emotion. Our brains are wired to fire up and take action when our emotions are aroused. People will only respond to a story if they care about the message. They must feel something. Knowledge of this fact can be used for good or ill, however. If you rile emotions to promote decisions that hurt people, you manipulate emotions for your benefit, not theirs. Out of shameful self-interest, many individuals and organizations promote outright lies to incite emotions in the service of their own agendas.

The best way to get people to care is to connect the story to something they already care about. When I read *Made to Stick*, I learned about a hospital that wanted to improve its workflow. Its administrators hired the design firm IDEO to help them convince the hospital staff that improvement was necessary. To do this, IDEO made a video that was shot from the perspective of a patient who enters the emergency room with a fractured leg. When hospital workers watched this video, they experienced what it was like to be a patient helplessly caught in a dysfunctional workflow, moved from place to place, waiting for long periods, all the while in pain. For the first time they saw hospital workflow from a different perspective, through the eyes of their patients, and they were suddenly able to care.

Videos are not always needed to help people connect with stories emotionally. Sometimes emotional connection can be accomplished by doing nothing more than showing the photograph of a real person who's involved in the story, which puts a face on an otherwise abstract problem. Sometimes it can be done by showing a graph that reveals the extent of the problem in simple and straight-forward terms. Bill Gates, the founder of Microsoft, first decided to focus many of the Bill and Melinda Gates Foundation's philanthropic efforts on solving world health problems when he saw a graph on the topic in the *New York Times* that caught his eye and touched his heart. More than anything else, however, we can help others care about our stories if we care about them ourselves. Nothing is more powerful than genuine passion.

Actionable

Despite my reservations about the term "actionable" because of the marketing hype surrounding it, I can't come up with another single word that says what I want to say better here. Effective stories make it easy for people to respond—to take action. Stories do this by suggesting, either directly or subtly, one or more ways to respond. If you want your stories to have an effect, you must suggest practical ways to build a bridge between the world that is and the world you desire.

Sequential

Narrative unfolds in a serial fashion, with a beginning, a middle, and an end. Stories are usually told sequentially by revealing facts at the proper moment. You might hint at things to come to build anticipation, but you should reveal a story's parts in a way that matches the chronology of events; builds concepts from simple ideas to more complex wholes; or states a thesis, validates it with evidence, and connects it to the audience's values, one point at a time in logical order.

When you tell a story verbally with the assistance of visuals, those visuals should not be revealed until the moment when they're relevant. As you refer to parts of a visual image, do something to highlight that part, such as pointing to it or visibly emphasizing it on the screen. Your audience should never struggle to locate what you're talking about. Whether you deliver the narrative verbally or in writing, if you are presenting complex concepts, build them piece by piece, adding complexity in stages as each previous stage is absorbed by the audience.

Despite their many problems, slideware products like PowerPoint and Keynote can be used quite effectively for statistical narrative. Slides are appropriate when verbal presentations can be enhanced through visual images. The sequential nature of slides, transitions, and the sudden appearance of words and images often pairs nicely with storytelling. Although I share Edward Tufte's opinion that PowerPoint encourages horrible presentations, I believe that its limitations and counterproductive recommendations can be bypassed to produce compelling stories. Tufte argues:

> With information quickly appearing and disappearing, the slide transition is an event that attracts attention to the presentation's compositional methods. Slides serve up small chunks of promptly vanishing information in a restless one-way sequence. It is not a contemplative analytical method; it is like television, or a movie with over-frequent random jump-cuts.[5]

5. Edward Tufte (2006) *Beautiful Evidence*. Graphics Press, page 160.

In fact, slideware in no way forces us to show only tiny fragments of information at a time or to venture forward and never backward. Tufte further asserts: "The rigid slide-by-slide hierarchies, indifferent to content, slice and dice the evidence into arbitrary compartments, producing an anti-narrative with choppy continuity."[6] But slides are no more prisoners of a hierarchy than pages of a book. Used properly, slides can complement the words of a gifted speaker with images that bring the story to life and express it clearly.

6. *ibid.* page 166

Slideware certainly isn't the only way to tell a quantitative story sequentially.

When using a medium that doesn't enforce a sequence by dividing content into slides, pages, or chapters, you must find another way to make the sequence obvious. For instance, you could tell a quantitative story on a single large screen or sheet of poster-sized paper. In such a case, you could choose from several methods to guide the reader through the information in the proper order. Comic books and graphic novels sequence narrative into panels arranged from left to right, top to bottom. A poster can number sections of the display or use arrows to reveal the suggested path.

Stories in the Wings

Hiding in the wings are more quantitative stories than we will ever know, waiting for their moments of truth. We must try to find them, coax them from the shadows, lead them to the stage, and give them a voice. You can apply these storytelling skills to make your own part of the world a better place.

14 THE INTERPLAY OF STANDARDS AND INNOVATION

When you design tables and graphs, you face many choices. Of the available alternatives, some are bad, some are good, some are best, and others are simply a matter of preference among equally good choices. By developing and following standards for the visual display of quantitative information, you can eliminate all but good choices once and for all. This dramatically reduces the time it takes to produce tables and graphs as well as the time required by your readers to make good use of them. Doing this will free up time to put your creativity to use where it's most needed.

Ask any of my friends, and they will tell you, without hesitation, that I have little respect for authority. Please don't misunderstand. I'm not living under an assumed name or hiding from the law. I simply don't like being told what to do. I prefer to examine things myself, get to know them, and then make my own informed decisions. Consequently, I'm not one who advocates rules for rules' sake. The term "standards" often leaves a bitter taste in my mouth. I don't like anything that gets in the way of doing good work and doing it efficiently. Nevertheless, through the course of my professional life, I've come to appreciate standards when they grow out of genuine need for direction, address only what must be addressed, and save me time. Standards that reside in huge tomes generally remain on the shelf gathering dust. Standards that actually get used are concise, easy to learn, easy to follow, and undeniably beneficial.

Ralph Waldo Emerson, one of the great practical philosophers of all time, wrote in his essay on self-reliance: "A foolish consistency is the hobgoblin of little minds."[1] This is often misquoted by omitting the word "foolish." Not all consistent practices are foolish. There are wise consistencies as well. These are standards that we follow because they produce the best result. These are practices that free us from redundant choices, conserving our energy and intelligence for the most deserving tasks. A good set of standards for the design of tables and graphs provides a framework for innovation in the face of important communication challenges. Standards and innovation are partners in the pursuit of excellence.

I'm amazed at how much time we waste making the same decisions over and over again simply because we've never taken the time to determine what works best for routine tasks, thereby making the decisions once and for all. Kevin Mullet and Darrel Sano beautifully express the benefits of a good set of standards:

> *Without minimizing the value of intuition as a problem solving tool, we propose that systematic design programs are more valuable from a communication standpoint than are ad hoc solutions; that intention is preferable to accident; that principled rationale provides a more compelling basis for design decisions than personal creative impulse.*[2]

1. Ralph Waldo Emerson (1841) *Essays: First Series, "Self-Reliance."*

2. Kevin Mullet and Darrel Sano (1995) *Designing Visual Interfaces.* Sun Microsystems, Inc., page 15.

You will have plenty of worthwhile opportunities for creativity as you work to understand numbers and present their message in a way that leads your readers to insight and action. "Design professionals learn quickly that constraints free the designer to focus . . . resources on those portions of the problem where innovation is most likely to lead to a successful product."[3] Your time is too valuable to waste. Keep your mind alert with challenging and productive activity. Don't dull it with redundant, meaningless choices.

3. Kevin Mullet and Darrel Sano (1995) *Designing Visual Interfaces*. Sun Microsystems, Inc., page 227.

Not only do wise consistencies save you time and effort, they do the same for your readers. Rather than calling readers' attention to ever-changing diversity in the design of tables and graphs, you can consistently use best practices to create a familiar form that invites the information to tell its story and engage their interest.

Think for a moment about standards that you've encountered in your own work. When have they worked? When have you ignored them and why? Here are some of the qualities that design standards for tables and graphs should exhibit:

- They should grow out of real needs and never grow beyond them.
- They should represent the most effective design practices.
- They should have effective communication as their primary objective.
- They should evolve freely as conditions change, or new insights are gained.
- They should be easy to learn and remember.
- They should save time.
- They should be applied to the software that is used to produce tables and graphs in the form of specific instructions and design templates.
- They should never inhibit creativity that actually improves communication.

I won't write your design standards for you. If I did, they would likely end up gathering dust on that shelf with all the others. The standards that will actually get used will be your own, the product of your own sweat, the rewards of your own labor. Whatever you do is worth doing well. Give a clear voice to the numbers that tell the story of your organization and clear visibility to the opportunities those numbers reveal. Produce tables and graphs that communicate.

Appendix A TABLE AND GRAPH DESIGN AT A GLANCE

Fundamental Steps in the Design Process

1. Determine your message.
2. Select the best means to display your message.
3. Design the display to show the information simply, clearly, and accurately.
   - Make the data (versus non-data) prominent and clear.
   - Remove all components that aren't necessary (both data and non-data components).
   - Reduce the visual salience of the remaining non-data components in comparison to the data.
   - Highlight the information that's most important to your message.

When to Use Tables vs. Graphs

Use Tables When	Use Graphs When
1. The display will be used to look up individual values	1. The message is contained in patterns, trends, and exceptions
2. The display will be only be used to compare individual values rather than whole series of values	2. Entire series of values must be seen as a whole and/or compared
3. Precise values are required	
4. Values involve more than one unit of measure	
5. Values must be presented at various levels of aggregation (i.e., summary and detail)	

Tables: Matching Relationships and Structural Types

Quantitative-to-Categorical Relationships

	Structural Type	
Relationship	**Unidirectional**	**Bidirectional**
Between a single set of quantitative values and a single set of categorical items	Yes	Not applicable because there is only one set of categorical items
Between a single set of quantitative values and the intersection of multiple categories	**Yes.** Sometimes this structure is preferable because of convention.	**Yes.** This structure saves space.
Between a single set of quantitative values and the intersection of multiple hierarchical categories	**Yes.** This structure can clearly display the hierarchical relationship by placing the separate levels of the hierarchy side by side in adjacent columns.	Yes. However, this structure does not display the hierarchy as clearly if its separate levels are split between the columns and rows.

Quantitative-to-Quantitative Relationships

	Structural Type	
Relationship	**Unidirectional**	**Bidirectional**
Among a single set of quantitative values associated with multiple categorical items	Yes	**Yes.** This structure works especially well because the quantitative values are arranged closely together for easy comparison.
Among distinct sets of quantitative values associated with a single categorical item	Yes	Yes. However, this structure tends to get messy as you add multiple sets of quantitative values.

Graph Selection Matrix

Featured Relationship	Value-Encoding Objects			
	Points (Scatter Plot / Dot Plot / Strip Plot)	**Lines** (Line Graph / Line Graph with Points)	**Bars** (Bar Graph vertical / Bar Graph horizontal)	**Boxes** (Box Plot vertical / Box Plot horizontal)
Time Series Values display how something changed through time (yearly, monthly, etc.)	Yes (as a *dot plot,* when you don't have a value for every interval of time)	Yes (to feature overall trends and patterns and support their comparisons)	Yes (vertical bars only, to feature individual values and to support their comparisons)	Yes (vertical boxes only, to display how a distribution changes through time)
Ranking Values are ordered by size (descending or ascending)	Yes (as a *dot plot,* especially when the quantitative scale does not begin at zero)	No	Yes	Yes (to display a ranked set of distributions)
Part-to-Whole Values represent parts (proportions) of a whole (for example, regional portions of total sales)	No	Yes (to display how parts of a whole have changed through time)	Yes	No
Deviation The difference between two sets of values (for example, the variance between actual and budgeted expenses)	Yes (as a *dot plot,* especially when the quantitative scale does not begin at zero)	Yes (when also featuring a time series)	Yes	No
Distribution Counts of values per interval from lowest to highest (for example, counts of people by age intervals of 10 years each)	Yes (as a *strip plot* to feature individual values)	Yes (as a *frequency polygon,* to feature the overall shape of the distribution)	Yes	Yes (when comparing multiple distributions)
Correlation Comparison of two paired sets of values (for example, the heights and weights of several people) to determine if there is a relationship between them	Yes (as a *scatter plot*)	No	Yes (as a *table lens,* especially when your audience is not familiar with *scatter plots*)	No
Geospatial Values are displayed on a map to show their location	Yes (as bubbles of various sizes on a map)	Yes (to display routes on a map)	No	No
Nominal Comparison A simple comparison of values for a set of unordered items (for example, products, or regions)	Yes (as a *dot plot,* especially when the quantitative scale does not begin at zero)	No	Yes	No

Appendix B   RECOMMENDED READING

The following books are some of the best that have been written about data visualization:

Edward Tufte is one of the world's leading theorists in the field of data visualization. All four of his books are filled with marvelous insights.

The Visual Display of Quantitative Information, Second Edition (2001), Graphics Press.

Envisioning Information (1990), Graphics Press.

Visual Explanations (1997), Graphics Press.

Beautiful Evidence (2006), Graphics Press.

William Cleveland has written two of the best guidelines for sophisticated graphing techniques, especially for statisticians.

The Elements of Graphing Data (1994), Hobart Press.

Visualizing Data (1993), Hobart Press.

Naomi Robbins has written a simple and practical book that makes many of the principles taught by Cleveland accessible to non-statisticians.

Creating More Effective Graphs (2005), John Wiley & Sons, Inc.

Robert Harris has written a comprehensive encyclopedia of charts in their many forms. This is a great reference to keep handy.

Information Graphics: A Comprehensive Illustrated Reference (1999), Oxford University Press.

Colin Ware has written the most accessible and comprehensive explanations that I've found about visual perception and how our current understanding of it can be applied to the presentation of information.

Visual Thinking for Design (2008), Elsevier, Inc.

Information Visualization: Perception for Design, Second Edition (2004), Morgan Kaufmann Publishers.

Robert Horn tackles the entire realm of visual thinking and asserts that a completely new visual language has emerged.

Visual Language Global Communication for the 21st Century (1998), MacroVU Press.

Stuart Card, Jock Mackinlay, and Ben Shneiderman provide the best introduction to research in the field of information visualization.

Readings in Information Visualization: Using Vision to Think (1999), Morgan Kaufmann.

My other two books teach principles and practices for presenting data in the form of dashboards and graphical techniques for exploring and analyzing data.

Information Dashboard Design: The Effective Visual Communication of Data (2006), O'Reilly Media, Inc.

Now You See It: Simple Visualization Techniques for Quantitative Analysis (2009), Analytics Press.

Although not about data visualization in particular, the following books address critical related topics:

Jonathan Koomey provides thoughtful instruction in the use and presentation of quantitative information in ***Turning Numbers into Knowledge: Mastering the Art of Problem Solving***, Second Edition (2008), Analytics Press.

Derrick Niederman and David Boyum provide a thoroughly readable and entertaining introduction to quantitative information as it is used in the world today in ***What the Numbers Say: A Field Guide to Mastering Our Numeric World*** (2003), Broadway Books.

Karen Berman and Joseph Knight provide a practical and accessible introduction to financial data—what people look for in it, how they use it, and why—in ***Financial Intelligence: A Manager's Guide to Knowing What the Numbers Really Mean*** (2006), Harvard Business Press.

Chip Heath and Dan Heath present clear and simple principles for presenting ideas in a way that people can understand, remember, care about, and respond to in ***Made to Stick: Why Some Ideas Survive and Others Die*** (2007), Random House.

Karen A. Schriver provides thorough guidelines for general document design in ***Dynamics in Document Design*** (1997), John Wiley & Sons, Inc.

If you're interested in geographical displays of quantitative data, I've found the following three books especially helpful:

Andy Mitchell, ***The ESRI Guide to GIS Analysis, Volume 1: Geographic Patterns and Relationships*** (1999), ESRI Press.

Cynthia Brewer, ***Designing Better Maps: A Guide for GIS Users*** (2005), ESRI Press.

Gretchen Peterson, ***GIS Cartography: A Guide to Effective Map Design*** (2009), CRC Press.

If you're responsible for presenting information in meetings using slideware such as PowerPoint or Keynote, the following books are filled with useful guidelines and inspiration:

Garr Reynolds, ***Presentation Zen: Simple Ideas on Presentation Design and Delivery*** (2008) and ***Presentation Zen Design: Simple Design Principles and Techniques to Enhance Your Presentations*** (2010), both published by New Riders.

Nancy Duarte, ***Slide:ology: The Art and Science of Great Presentations*** (2008), O'Reilly Media, Inc., and ***Resonate: Present Visual Stories that Transform Audiences*** (2010), John Wiley and Sons, Inc.

Cliff Atkinson, ***Beyond Bullet Points: Using Microsoft PowerPoint to Create Presentations that Inform, Motivate, and Inspire***, Third Edition (2011), Microsoft Press.

Jerry Weisman, ***Presenting to Win: The Art of Telling Your Story*** (2009), Pearson Education Inc.

Appendix C   ADJUSTING FOR INFLATION

When the value of money decreases over time, we refer to this as inflation. To meaningfully and accurately compare money in the past to money today, you must express both using an equal measure of value. Doing this is called adjusting for inflation. Because most readers of this book live in the United States of America, I'll use U.S. dollars to illustrate how this is done. You can adjust for inflation using any of the following approaches:

- Convert today's dollars into their equivalent in a past year that is relevant to your purpose; or
- Convert dollars from the past into their equivalent in today's dollars; or
- Convert both today's dollars and dollars from past years into the same measure of value as of some other point in time (e.g., dollars in the year 2000).

If you're not in the habit of doing this, don't get nervous. It takes a little extra work but isn't difficult. The process requires the use of an *inflation index*. Several are available, but the two that are most commonly used in the United States are the Consumer Price Index (CPI), published by the Bureau of Labor Statistics (BLS), and the Gross Domestic Product (GDP) deflator, published by the Bureau of Economic Analysis (BEA). I'll use the CPI to illustrate the process. The CPI represents the dollar in terms of its buying power relative to goods that are typically purchased by consumers (food, utilities, etc.). CPI values are researched and computed for a variety of representative people, places, and categories of consumer goods. Let's look at a version of the CPI that represents an average of all classes of people across all U.S. cities purchasing all types of goods for the years 1990 through 2010:

Year	Jan	Feb	Mar	Apr	May	Jun	Jul	Aug	Sep	Oct	Nov	Dec	Annual
1990	127.4	128.0	128.7	128.9	129.2	129.9	130.4	131.6	132.7	133.5	133.8	133.8	130.7
1991	134.6	134.8	135.0	135.2	135.6	136.0	136.2	136.6	137.2	137.4	137.8	137.9	136.2
1992	138.1	138.6	139.3	139.5	139.7	140.2	140.5	140.9	141.3	141.8	142.0	141.9	140.3
1993	142.6	143.1	143.6	144.0	144.2	144.4	144.4	144.8	145.1	145.7	145.8	145.8	144.5
1994	146.2	146.7	147.2	147.4	147.5	148.0	148.4	149.0	149.4	149.5	149.7	149.7	148.2
1995	150.3	150.9	151.4	151.9	152.2	152.5	152.5	152.9	153.2	153.7	153.6	153.5	152.4
1996	154.4	154.9	155.7	156.3	156.6	156.7	157.0	157.3	157.8	158.3	158.6	158.6	156.9
1997	159.1	159.6	160.0	160.2	160.1	160.3	160.5	160.8	161.2	161.6	161.5	161.3	160.5
1998	161.6	161.9	162.2	162.5	162.8	163.0	163.2	163.4	163.6	164.0	164.0	163.9	163.0
1999	164.3	164.5	165.0	166.2	166.2	166.2	166.7	167.1	167.9	168.2	168.3	168.3	166.6
2000	168.8	169.8	171.2	171.3	171.5	172.4	172.8	172.8	173.7	174.0	174.1	174.0	172.2
2001	175.1	175.8	176.2	176.9	177.7	178.0	177.5	177.5	178.3	177.7	177.4	176.7	177.1
2002	177.1	177.8	178.8	179.8	179.8	179.9	180.1	180.7	181.0	181.3	181.3	180.9	179.9
2003	181.7	183.1	184.2	183.8	183.5	183.7	183.9	184.6	185.2	185.0	184.5	184.3	184.0
2004	185.2	186.2	187.4	188.0	189.1	189.7	189.4	189.5	189.9	190.9	191.0	190.3	188.9
2005	190.7	191.8	193.3	194.6	194.4	194.5	195.4	196.4	198.8	199.2	197.6	196.8	195.3
2006	198.3	198.7	199.8	201.5	202.5	202.9	203.5	203.9	202.9	201.8	201.5	201.8	201.6
2007	202.4	203.5	205.4	206.7	207.9	208.4	208.3	207.9	208.5	208.9	210.2	210.0	207.3
2008	211.1	211.7	213.5	214.8	216.6	218.8	220.0	219.1	218.8	216.6	212.4	210.2	215.3
2009	211.1	212.2	212.7	213.2	213.9	215.7	215.4	215.8	216.0	216.2	216.3	215.9	214.5
2010	216.7	216.7	217.6	218.0	218.2	218.0	218.0	218.3	218.4	218.7	218.8	219.2	218.1

If you need inflation values that are more directly tied to a specific item (e.g., food), a particular class of person (e.g., clerical workers), or a particular part of the country (e.g., the San Francisco area), it's likely that you'll find them in an index. Visit the Bureau of Labor Statistics' (BLS) website and select what you need from the broad range of available data. It's easy to download data from the website to your own computer and read it using software such as Excel. I was able to get the information for the sample table simply by copying it from the website into my clipboard and pasting it into Excel.

Once you have an index in a form that's suited for your use (e.g., in a spreadsheet), here's how you actually use it. The current version of the CPI uses the value of dollars from 1982 to 1984 as its baseline. Each value in the index represents the value of the dollar at that time compared to its value from 1982 to 1984. For instance, according to the previous table, in January of 1990, the value of the dollar was 127.4% of its value in 1982 to 1984, and for the year 1990 as a whole, it was 130.7% of its value in 1982 to 1984. Typically, if you were comparing money across a range of time, you would express everything according to the value of money at some point along that range, and most often in terms of its value at the point in time that you're producing your report. If you produced your report in 2010, including values ranging from 1998 to 2010, you would likely want to convert all the values to their 2010 equivalent. Here's how you would convert a year 1998-value of $100,000 into its year-2010 equivalent, assuming that you are only dealing with one value per year (e.g., as opposed to monthly or quarterly values):

1. Find the index value for the year 2010, which is 218.1.
2. Find the index value for the year 1998, which is 163.0.
3. Divide the index value for 2010 by the index value for 1998, which results in 1.338037.
4. Multiply the dollar value for 1998, $100,000, by the results of step 3, 1.338037, which results in $133,803.70. If you wish, you can round to the nearest whole dollar, producing the final result of $133,804.

Because year 2010 dollars are already expressed as 2010 dollars, you don't have to convert them. If you're using spreadsheet software, such as Excel, setting up the formulas to convert money using an inflation index like the CPI is quite easy to do.

Whether you decide to express money across time using an inflation index to convert it to a common base or to use the actual values without adjusting them for inflation, you should always clearly indicate on your report what you've done. Don't leave your readers guessing. As a communicator of important information, you should get into the habit of identifying the way that you've expressed the value of money. If you haven't adjusted for inflation, you can simply state somewhere on the report that you are using Current U.S. Dollars. If you have adjusted for inflation, use a statement like "Adjusted to a base of year 2010 U.S. dollars or Adjusted according to the CPI using a baseline of year 2010."

For additional information on this topic, along with further instruction in the use of quantitative information, I recommend that you get a copy of Jonathan Koomey's excellent book *Turning Numbers into Knowledge*, published by Analytics Press.

Appendix D — CONSTRUCTING TABLE LENS DISPLAYS IN EXCEL

A *table lens* display provides a way to view potential correlations between several quantitative variables simultaneously. The example below displays four quantitative variables—profit, sales, shipping costs, and discount percentage—for 25 states.

With the states sorted by profit in descending order, we can use this table lens display to see whether any other columns of bars (sales, etc.) reveal either a corresponding high to low value pattern, indicating a positive correlation, or an opposite low to high value pattern, indicating a negative correlation.

Table lens displays are sometimes useful even when only two quantitative variables must be compared. Because a table lens uses a series of bar graphs, people who are not familiar with scatter plots can often find it understandable.

Even though Excel does not provide a chart type called a table lens, one like the example above can be constructed with relative ease. To do so, follow these steps:

- Enter the categorical labels for the data set (e.g., states in the example above) into a single column of the spreadsheet.
- Enter each set of quantitative values into a separate column (e.g., profit, sales, and so on in the example above), with the one that you want to feature most in the column immediately to the right of the column of labels.
- Sort the rows by the first column of quantitative values (profit in the example above) in descending order.

These steps were written based on my version of Excel 2010. This can be done in earlier versions of Excel as well but might involve slightly different menu selections.

- Create a horizontal bar graph that consists only of the labels and the first column of quantitative values.

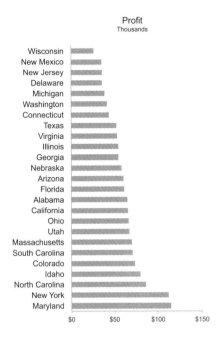

- Click on the Y axis to select it, then right click and choose *Format Axis*. Check the *Categories in reverse order* checkbox. While still in the pop-up menu, select *Line* and set *Color* to *No Line*.

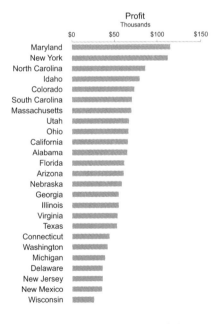

- Adjust the size, shape, and visual attributes of this graph as necessary. In a moment we'll make copies of this graph to display the other variables, so you'll want to make all of your design changes here.

- Click on the plot area of the graph to select it. Now click on the resize handle on the left side of the plot area (next to the categorical labels) and move it to the right or left just a bit. When you manually resize the plot area in this manner, Excel will lock its size from automatically adjusting, which will be useful when you hide the categorical labels in the copies you're about to make.
- Copy and paste the graph once for each additional column of quantitative values, and place the copies side by side to the right of the first graph.
- For each of the new graphs, right click on it and choose *Select Data*, which allows you to change the set of quantitative values that it contains to include a different column of quantitative values.
- In each graph except for the first, select the Y axis to highlight it and press *Delete* to remove the labels. Only the leftmost graph needs to display the labels.
- Position the graphs close to each other, side by side, with the rows of bars aligned with one another.

The result should be a series of graphs similar the example that we began with on the first page of this appendix.

Appendix E   CONSTRUCTING BOX PLOTS IN EXCEL

Microsoft Excel does not support direct construction of box plots, but you can play some tricks with its charting features to construct graphs that look like box plots. Simple 3-value box plots (top of box, center, and bottom of box) and slightly more complex 5-value box plots (top of line, top of box, center, bottom of box, and bottom of line) can both be made using Excel. The following instructions were written based on my version of Excel 2010. Box plots can also be constructed using this method in earlier versions of Excel, but the menu selections might vary slightly depending on your version and platform.

The first step is to enter data values for each box. The following two tables illustrate how you would enter data for a three-value box plot (left) and a five-value box plot (right).

Values	Box 1	Box 2	Box 3	Box 4
Top of Box	434	343	433	344
Center	350	300	360	280
Bottom of Box	273	252	290	225

Values	Box 1	Box 2	Box 3	Box 4
Top of Box	400	325	390	325
Top of Line	434	343	433	344
Center	350	300	360	280
Bottom of Line	273	252	290	225
Bottom of Box	320	280	330	250

Although these examples contain data about distributions of four sets of values, one box each, you may have any number of them. Typically, in this version of a three-value box plot, the top of the box would represents the highest value, the bottom would represent the lowest value, and the center would represent the median of the data set. In this version of a five-value box plot the top of the line would represent the highest value, the bottom of the line would represent the lowest value, the center of the box would represent the median, the top of the box would represent the 75th percentile, and the bottom of the box would represent the 25th percentile.

Here's a preview of how these two box plots will look when we've worked through all of the steps:

Note the order of the values in the five-value table. Even though it would make more sense to place the top of line value at the top, because it's the highest, and the bottom of line value at the bottom, because it's the lowest, for the box plot to work when constructed in this way, you must enter the values in the order shown.

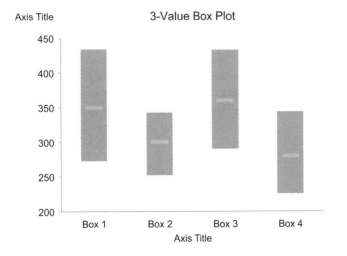

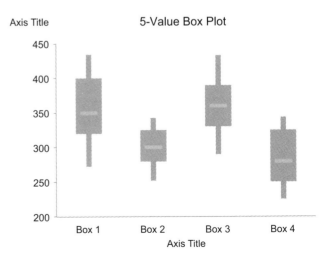

Using a table of either 3 or 5 values as illustrated above, create a line graph based on the table, with points on the line. The result will look like this:

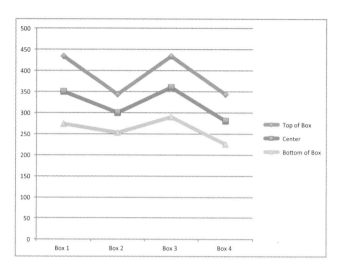
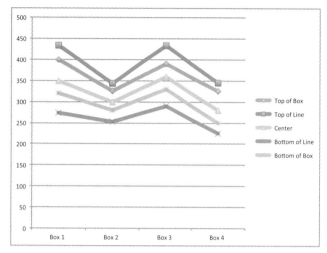

Do each of the following steps:

- Remove the border around the graph.
- Remove or lighten the grid lines.
- Remove the legend.
- Remove the tick marks along the X axis.
- Select the line in the middle, right-click to access the pop-up menu, and then select *Format Data Series*.
  - From the Options tab turn on *Up-down* bars.
  - For the 5-value box plot only, turn on *High-low lines* also.

The result will look like this:

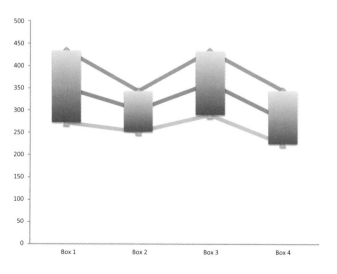
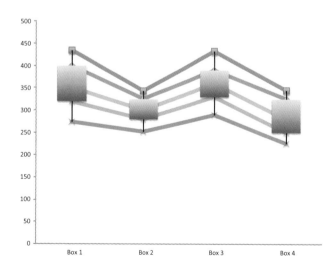

Adjust the scale on the Y axis to begin a little below the lowest value and end a little above the highest value, as illustrated on the following page:

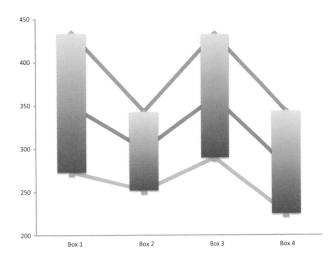

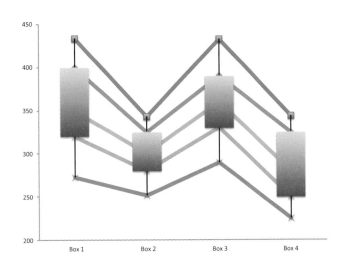

Select each of the lines except for the one in the middle, and do the following to each:

- Right-click to access the pop-up menu and then select *Format Data Series*.
- In the *Options* tab, adjust the gap width to 100%.
- In the *Marker Style* tab, change the Color setting to *No marker*.
- In the *Line* tab, change the *Color* setting to *No line*.

At this point the graphs should look something like this:

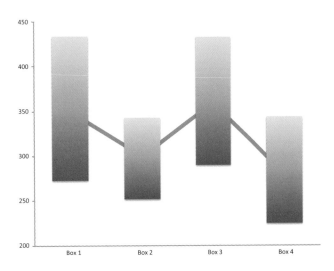

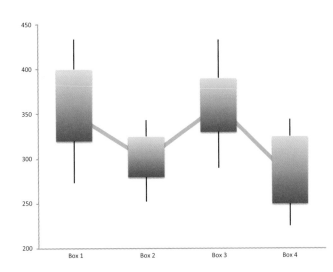

- Select the *up-down bars*, right-click to access the pop-up menu, and select *Format Down Bars*.
- In the *Fill* tab set the *Color* to one that is lighter than the default of black.
- In the *Shadow* tab, deselect *Shadow*.

For the 5-value box plot only, do the following:

- Select the *high-low lines*, right-click to access the pop-up menu, and select *Format High-Low Lines*.
- In the *Line* tab set the *Color* to the one that matches the box.
- In the *Line* tab, at the top select *Weights and Arrows* and set the weight of the line to one that is thick enough to easily see.

Select the remaining line (the one that was originally in the middle), right-click the pop-up menu, and select *Format Data Series*.

- In the *Marker Style* tab select the long dash and set it to a larger size.
- In the *Marker Fill* tab set the *Color* to one that is darker than the box.
- In the *Marker Line* tab, set *Color* to *No line*.
- In the *Line* tab, set *Color* to *No line*.

The result will look like this:

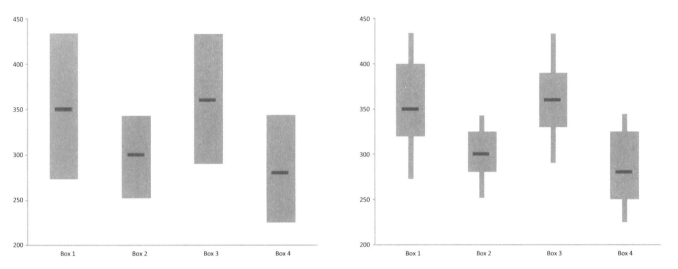

To complete the box plot, add a chart title and titles for the axes, as needed, remove the unnecessary tick marks along the X axis, and polish it up to look exactly as you wish.

Appendix F   ANSWERS TO PRACTICE IN
SELECTING TABLES AND GRAPHS

Scenario #1

1. **Table or graph?** Because the CFO needs a simple reference that contains budgeted and actual headcount and expense figures for the current quarter so that she can quickly locate the numbers for each of her meetings, the report will be used primarily to look up data. Therefore, a table will work most effectively.

2. **If a table, which kind?** Assuming that there are more departments than there are sets of quantitative values (i.e., more than four), it would work best to display one row for each department and separate columns for the four sets of quantitative values.

3. **If a graph, what kind of relationship?** Not applicable.

4. **If a graph, which graphical objects for quantitative encoding?** Not applicable.

5. **Anything else?** Nothing of significance comes to mind.

Scenario #2

1. **Table or graph?** Because you need to show a pattern rather than individual values, a graph will work best.

2. **If a table, which kind?** Not applicable.

3. **If a graph, what kind of relationship?** This graph needs to display two relationships of equal significance to your message: 1) a part-to-whole relationship, comparing the relative revenue contributions to total revenue of each of the product lines, and 2) a time-series relationship, showing how the relative contributions of the individual product lines have changed during the past five years.

4. **If a graph, which graphical objects for quantitative encoding?** Although part-to-whole relationships are usually encoded as bars, we need to show five sets of values, one for each product line, so the trend of revenue contribution across time would be difficult to see using five sets of five bars, with a distinct bar for each product line and a set of bars for each year. A distinct line on the graph to represent each of the product lines, each line a different color, will display the trend and relative values of the product lines more clearly.

5. **Anything else?** To clearly emphasize the decline over time of the programming and utilities product lines, you might want to make their lines stand out somewhat from the others, such as by increasing their thickness.

Scenario #3

1. **Table or graph?** If your director was comfortable interpreting measures of average like means and medians and measures of variation like standard deviations, you could provide him with a very simple table containing the median, mean, and standard deviation for each of the two classes. Because he's not comfortable with standard deviations, you should use an approach that displays the information graphically.
2. **If a table, which kind?** Not applicable.
3. **If a graph, what kind of relationship?** The shape of the distribution of the ratings across the five values (1 through 5) can both be displayed by means of a frequency distribution.
4. **If a graph, which graphical objects for quantitative encoding?** Because you want to display the shape of the distribution, a histogram using five vertical bars (one per value) or a frequency polygon using a line could be used.
5. **Anything else?** Because you want to compare the ratings of two different courses, a frequency polygon with two lines (one per course) would support this slightly better than a pair of histograms (one per course).

Scenario #4

1. **Table or graph?** Even though you will be communicating a ranking of the four customer service centers to point out the ones that need the most improvement, your message consists of only four values. Given such a tiny data set, this message can be communicated more efficiently using a table.
2. **If a table, which kind?** This is a simple list, with one value for each customer service center.
3. **If a graph, what kind of relationship?** Not applicable.
4. **If a graph, which graphical objects for quantitative encoding?** Not applicable.
5. **Anything else?** The message will be clearest if you sort the customer service centers in the table in order of their average rating.

Scenario #5

1. **Table or graph?** You want to show a pattern, so a graph will definitely work best.
2. **If a table, which kind?** Not applicable.
3. **If a graph, what kind of relationship?** It will display a correlation between the number of workers and their productivity.
4. **If a graph, which graphical objects for quantitative encoding?** A scatter plot with points for each daily measure of headcount and productivity would show their correlation best. To highlight possible changes to the correlation since the new Operations Manager took over, you could assign different colors to the data points associated with the period before he arrived versus the period since.

5. **Anything else?** Perhaps data points for the period after the new Operations Manager began could vary in color intensity for each month (e.g., from light to dark red) to reveal whether the correlation increased or decreased over time.

Scenario #6

1. **Table or graph?** A graph will highlight the dramatic nature of this information more effectively than a table would. The quick fall-off of the income contribution of donations to your organization from 87% for the top 10% of the donations to 1% for the bottom 50% lends itself to a visual display.

2. **If a table, which kind?** Not applicable.

3. **If a graph, what kind of relationship?** This scenario is a little tricky. Essentially, the information involves a part-to-whole relationship: the revenue contribution of each 10% group of donations as a percentage of total revenue. Even though our separation of the donations into 10 groups involved a ranking of them based on their revenue amount, the point of our message is the contribution of our biggest donations to overall income, compared to the contributions of the other groups of donations of decreasing size.

4. **If a graph, which graphical objects for quantitative encoding?** A part-to-whole relationship is most effectively displayed using bars. In this case each bar represents the percentage revenue contribution of each 10% group of donations.

5. **Anything else?** Very clear labeling along both axes is necessary to avoid confusion between the 10% groups of donations vs. the percentage contribution of each group to overall income. Also, a title that clearly and succinctly states the essential message, such as "The Top 10% of Our Donations Account for 87% of Our Income" would catch the reader's attention, allowing the graph to then dramatically reinforce the message.

Appendix G ANSWERS TO PRACTICE IN TABLE DESIGN

Exercise #1

Here's the original table:

Quarter-to-Date Sales Rep Performance Summary
Quarter 2, 2011 as of March 15, 2011

Sales Rep	Quota	Variance to Quota	% of Quota	Forecast	Actual Bookings
Albright, Gary	200,000	-16,062	92	205,000	183,938
Brown, Sheryll	150,000	84,983	157	260,000	234,983
Cartwright, Bonnie	100,000	-56,125	44	50,000	43,875
Caruthers, Michael	300,000	-25,125	92	324,000	274,875
Garibaldi, John	250,000	143,774	158	410,000	393,774
Girard, Jean	75,000	-48,117	36	50,000	26,883
Jone, Suzanne	140,000	-5,204	96	149,000	134,796
Larson, Terri	350,000	238,388	168	600,000	588,388
LeShan, George	200,000	-75,126	62	132,000	124,874
Levensen, Bernard	175,000	-9,267	95	193,000	165,733
Mulligan, Robert	225,000	34,383	115	275,000	259,383
Tetracelli, Sheila	50,000	-1,263	97	50,000	48,737
Woytisek, Gillian	190,000	-3,648	98	210,000	186,352

List each of the problems that you see in the design of this table:

1. The grid that delineates the columns and rows is far too heavy and isn't necessary at all.
2. The combined use of boldfacing and gray fill color to delineate the headers is excessive, grabbing too much attention relative to the values.
3. The center alignment of the headers fails to preview the alignment of their values.
4. The values in the % of Quota column aren't clear without percentage signs.
5. Nothing in the columns that contain dollar values indicates the unit of measure.
6. Overall performance for the entire group cannot be determined without summing the values in the columns.
7. Nothing has been done to make it easy to compare the performance of the individual sales representatives.
8. The most important and useful set of quantitative values is bookings, yet they are in the last column, farthest from the names of the sales representatives, so a great deal of eye movement is required to pair them.
9. The Variance to Quota and % of Quota columns appear before the Bookings column, which is used to calculate them.

Suggest a solution to each of these problems:

1. Eliminate the grid entirely, using only white space to delineate the columns and the rows.
2. Set the headers apart from the values using a rule line and nothing more. The headers could be boldfaced to set them apart further, but this really isn't necessary and would make them more prominent than the values they describe.
3. Align the headers consistently with the values they describe.
4. Format the values in the percentage column with percentage signs.
5. Format the summary values in the columns containing dollars to clarify the unit of measure, assuming that the context does not require further clarification to indicate U.S. dollars, Canadian dollars, etc.
6. Add column summaries to provide measures of overall performance.
7. To enhance comparison of sales representative performance, sort the rows by bookings, and add a column that expresses the bookings of each sales representative as a percentage of total bookings.
8. Place the column of bookings values immediately to the right of the sales representatives' names.
9. Place the calculated columns containing the bookings-to-quota variance and bookings percentage-of-quota values to the right of both the bookings and the quota columns.

Here's an example of what the table might look like when designed effectively:

Quarter-to-Date Sales Rep Performance Summary
Quarter 2, 2011 as of March 15, 2011

Sales Rep	Actual Bookings	% of Total Bookings	Forecasted Bookings	Quota	Bookings to Quota Variance	Bookings % of Quota
Larson, Terri	588,388	22.1%	600,000	350,000	238,388	168%
Garibaldi, John	393,774	14.8%	410,000	250,000	143,774	158%
Caruthers, Michael	274,875	10.3%	324,000	300,000	-25,125	92%
Mulligan, Robert	259,383	9.7%	275,000	225,000	34,383	115%
Brown, Sheryll	234,983	8.8%	260,000	150,000	84,983	157%
Woytisek, Gillian	186,352	7.0%	210,000	190,000	-3,648	98%
Albright, Gary	183,938	6.9%	205,000	200,000	-16,062	92%
Levensen, Bernard	165,733	6.2%	193,000	175,000	-9,267	95%
Jone, Suzanne	134,796	5.1%	149,000	140,000	-5,204	96%
LeShan, George	124,874	4.7%	132,000	200,000	-75,126	62%
Tetracelli, Sheila	48,737	1.8%	50,000	50,000	-1,263	97%
Cartwright, Bonnie	43,875	1.6%	50,000	100,000	-56,125	44%
Girard, Jean	26,883	1.0%	50,000	75,000	-48,117	36%
Total	$2,666,591	100.0%	$2,908,000	$2,405,000	$261,591	111%

Exercise #2

Here's the original table:

Mortgage Loan Rates
Effective September 1, 2011

Loan Type	Term	Points	Lender	Rate
Adjustable	15	0	ABC Mortgage	6.0%
Adjustable	15	0	BCD Mortgage	6.0%
Adjustable	15	0	CDE Mortgage	6.0%
Fixed	15	0	ABC Mortgage	6.25%
Fixed	15	0	BCD Mortgage	6.75%
Fixed	15	0	CDE Mortgage	7.0%
Adjustable	30	.5	ABC Mortgage	6.125%
Adjustable	30	.5	BCD Mortgage	6.25%
Adjustable	30	.5	CDE Mortgage	6.5%
Fixed	30	.5	ABC Mortgage	6.5%
Fixed	30	.5	BCD Mortgage	7.0%
Fixed	30	.5	CDE Mortgage	7.25%
Adjustable	15	1	ABC Mortgage	5.675%
Adjustable	15	1	BCD Mortgage	5.675%
Adjustable	15	1	CDE Mortgage	5.75%
Fixed	30	1	ABC Mortgage	6.5%
Fixed	30	1	BCD Mortgage	6.5%
Fixed	30	1	CDE Mortgage	7.0%
Adjustable	15	1	ABC Mortgage	5.675%
Adjustable	15	1	BCD Mortgage	5.675%

List each of the problems that you see in the design of this table:

1. The information is not grouped and sorted properly for looking up all the rates for a single lender.
2. The Points and the Lender columns are not adequately delineated. There is not enough white space to view them separately without effort.
3. The fluctuating precision of the values in the Points and Rate columns makes them difficult to read when scanning vertically.
4. The row headers (i.e., values in the Loan Type, Term, and Points columns) are unnecessarily repeated on each row, making it difficult to see when one group ends and the next begins.

Suggest a solution to each of these problems:

1. Place the lender names in the first column, followed by the other categorical values, with the rates in the right column.
2. Sometimes when a column with right-aligned values is positioned immediately to the left of a column with left-aligned values, there isn't enough white space between them to separate them adequately. This can be corrected with most software programs by placing a blank column between them to insert more white space.
3. Select the appropriate numeric precision; then, format all values to display this level of precision.
4. When rows are grouped by categorical values that appear to the left of the quantitative values, you only need to display those values on the first row of the group and on the first row of each new page that contains the group.

Here's an example of what the table might look like when designed effectively:

Mortgage Loan Rates by Lender Effective September 1, 2011				
Lender	**Loan Type**	**Term**	**Points**	**Rate**
ABC Mortgage	Adjustable	15 Yr	0.0	6.000%
			0.5	5.750%
			1.0	5.675%
		30 Yr	0.0	6.125%
			0.5	5.875%
			1.0	5.500%
	Fixed	15 Yr	0.0	6.250%
			0.5	6.000%
			1.0	5.750%
		30 Yr	0.0	6.500%
			0.5	6.000%
			1.0	5.750%
BCD Mortgage	Adjustable	15 Yr	0.0	6.000%
			0.5	5.750%
			1.0	5.675%
		30 Yr	0.0	6.125%

Note that in this example, the lenders' names have been highlighted using boldfacing. This isn't necessary, but because the purpose of the table is to look up specific lenders, it is useful to make the names stand out visually from the other data.

Exercise #3

Here's the original table:

2011 Marketing Department Expenses			
Quarter	**Transaction Date**	**Expense Type**	**Expense**
	9/28/2011	Software	3837.05
	9/28/2011	Computer Hardware	10873.34
	9/29/2011	Travel	2939.95
	9/30/2011	Supplies	27.53
Qtr 4	10/1/2011	Supplies	17.37
	10/1/2011	Postage	23.83
	10/3/2011	Computer Hardware	3948.85
	10/3/2011	Software	555.98
	10/3/2011	Furniture	739.37
	10/3/2011	Travel	28.83
	10/4/2011	Entertainment	173.91
	10/15/2011	Travel	33.57
	10/16/2011	Membership Fees	395.93
	10/16/2011	Conference Registration	2195.00

List each of the problems that you see in the design of this table:

1. The information is not conveniently arranged for the examination of expenses by expense type per quarter. The current arrangement would require a great deal of searching and summing to generate the needed information.

2. The table contains values that aren't necessary and that consequently just get in the way. Because the marketing manager doesn't need date information that is more detailed than the quarter level, the transaction dates are not only unnecessary, but they force the information down to a level of detail that increases its volume considerably.

3. Highlighting the quarters, expense types, and expense amounts through the use of boldfacing is unnecessary. Everything but the transaction dates has been highlighted to make the dates appear less important, but the dates are totally unnecessary and therefore shouldn't be included at all.

4. If the dates were needed, their fluctuating number of digits would render them inefficient to read.

5. Summary values at the required level of quarter and expense type are missing.

6. The expense amounts are missing the comma used to group every three whole number digits, making them harder than necessary to read.

7. Tying expense types and expenses closely together, if it were useful, could have been achieved through a better means of visually grouping this information. Even if a distinct font were appropriate, the font that was used in this table is not very legible.

8. The heavy vertical rules lead the eyes to scan downward through the columns, yet horizontal scanning between the expense types and expenses is the primary way that this table should be read.

9. Failing to repeat the quarter row header at the top of each new page could result in pages that have no row header at all, forcing readers to flip backward to previous pages to determine the quarter.

Suggest a solution to each of these problems:

1. Arrange the expense types in a column along the left with individual quarters across the top; this arrangement nicely supports the need to examine expenses by type in total and by quarter.

2. Eliminate the transaction dates entirely, displaying expenses only at the level of expense type and quarter.

3. By eliminating the transaction dates, you eliminate the need to make the other values stand out in relation to them. However, it would be useful to highlight total expenses by expense type as distinct and somewhat more important than the quarterly expenses.

4. Date formatting is no longer a problem because the dates have been eliminated.

5. When you eliminate the transaction dates, quantitative values will be summarized at the appropriate level of expense type by quarter. Total expenses by expense type can be added as a separate column.

6. Expense value formats can be corrected by adding the commas to group every three whole-number digits.

7. Eliminate the distinct font because it serves no useful purpose.

8. Eliminate the vertical rules so that horizontal scanning is not interrupted.

9. The elimination of the transaction dates eliminates the need to repeat row headers on each new page because there is now only one row per row header.

Here's an example of what the table might look like when designed effectively:

2011 Marketing Department Expenses

Expense Type	Total	Qtr 1	Qtr 2	Qtr 3	Qtr 4
Computer Hardware	12,970.40	3,883.64	8,352.83	0.00	733.93
Entertainment	4,778.52	736.94	1,873.03	185.01	1,983.54
Equipment	463.79	73.93	105.93	283.93	0.00
Furniture	2,541.11	0.00	108.83	493.83	1,938.45
Membership Fees	1,595.00	95.00	1,500.00	0.00	0.00
Postage	292.59	27.83	186.83	57.93	20.00
Supplies	444.06	117.93	75.39	74.82	175.92
Travel	12,674.09	5,938.93	3,978.39	863.02	1,893.75

Note that the Total column has been placed immediately to the right of the Expense Type column, making it very easy and efficient to associate the two sets of values. Because total expenses per expense type are somewhat more useful than any individual column of quarterly expenses, this arrangement of the columns makes the total slightly easier to read.

Exercise #4

Answers don't apply to this exercise.

Exercise #5

Here's one potential design solution to the proposed scenario:

YTD Product Performance through 5/31/11

Our top two products account for more than 89% of revenue and 95% of profit!

Product	Revenue Dollars (1,000s)	% of Total	Cum % of Total	Profit Dollars (1,000s)	% of Total	Cum % of Total
I	9,266	74.01%	74.01%	5,969	79.15%	79.15%
F	1,957	15.63%	89.64%	1,197	15.87%	95.02%
G	602	4.81%	94.45%	207	2.75%	97.77%
E	402	3.21%	97.66%	15	0.20%	97.97%
B	132	1.05%	98.71%	73	0.96%	98.93%
D	92	0.74%	99.45%	39	0.51%	99.44%
C	40	0.32%	99.77%	32	0.43%	99.88%
H	20	0.16%	99.93%	5	0.07%	99.94%
A	7	0.05%	99.98%	4	0.05%	99.99%
J	2	0.02%	100.00%	1	0.01%	100.00%
Total	$12,520	100.00%	100.00%	$7,541	100.00%	100.00%

Some of the highlights of this solution are:

1. The values have been sorted in descending order by revenue, thus placing the products in order of performance, beginning with the best. The values could also have been sorted by profit or by a combination of revenue and profit. Given this arrangement of values, with revenue first, then profit, it was appropriate to sort by revenue.

2. Because the only measures of performance that need to be communicated are revenue and profit, other measures that might have been included (e.g., cost or units sold) are left out.

3. The message that the top two products performed well above the rest has been highlighted in multiple ways: 1) boldfacing these two rows, 2) placing a concise textual statement of this fact in blue above the table, and 3) placing blue borders around the two measures that most directly present the contribution of these two products to overall revenue and profit.

4. Additional measures were calculated and included to enhance the meaning and clarity of the relative revenue and profit contributions of each product, including percent of total and cumulative percent of total. These measures make the limited contribution of the worst-performing products stand out even more than the dollars alone.

5. The title clearly states the period of time covered by the performance figures. Without this information, the meaning of the table would suffer even in the beginning and would quickly decrease in worth with the passage of time.

The design in this example solution is not the only design that would work effectively. Even if your design is significantly different, compare its merits to those of the example table to see how well it did in meeting the objectives specified in the scenario.

Exercise #6

Here's one potential design solution to the proposed scenario:

Sales Summary by Region (USD)
1st Quarter, 2011
Regions and Products are Sorted by Revenue
Data as of 02/07/11

Region	Revenue	% of Total Revenue	Expenses	Profit	% of Total Profit	Average Revenue per Salesperson	Avg Order Size	Coats	%	Skirts	%	Dresses	%	Pants	%	Shirts	%
								Product Revenue									
Europe	75,904,604	31.1%	40,988,486	34,916,118	22.3%	843,384	10,050	30,299,323	39.9%	14,135,203	18.6%	12,982,833	17.1%	10,384,302	13.7%	8,102,943	10.7%
Canada	51,572,694	21.1%	17,534,716	34,037,978	21.7%	991,783	9,468	17,938,444	34.8%	11,364,033	22.0%	9,253,400	17.9%	7,028,474	13.6%	5,988,343	11.6%
Western U.S.	42,660,178	17.5%	11,944,850	30,715,328	19.6%	1,185,005	9,276	12,503,954	29.3%	10,376,432	24.3%	8,503,942	19.9%	6,738,453	15.8%	4,537,397	10.6%
Eastern U.S.	33,977,385	13.9%	7,135,251	26,842,134	17.2%	1,415,724	8,263	8,640,293	25.4%	9,103,845	26.8%	7,270,982	21.4%	5,123,044	15.1%	3,839,221	11.3%
Central U.S.	26,139,598	10.7%	3,920,940	22,218,658	14.2%	1,742,640	8,885	4,239,443	16.2%	8,239,484	31.5%	6,039,461	23.1%	4,682,776	17.9%	2,938,434	11.2%
Asia	14,135,278	5.8%	6,360,875	7,774,403	5.0%	642,513	4,639	1,323,928	9.4%	1,938,843	13.7%	5,944,832	42.1%	2,543,343	18.0%	2,384,332	16.9%
Total (or Avg)	$244,389,737	100.0%	$87,885,118	$156,504,619	100.0%	$1,022,551	$8,823	$74,945,385	30.7%	$55,157,840	22.6%	$49,995,450	20.5%	$36,500,392	14.9%	$27,790,670	11.4%

Some of the highlights of this solution are:

1. Placing all measures for each region in a single row makes it easy to look up any value for a particular region.
2. Placing all measures for each region in a single row also makes comparisons of a single measure among all six regions easy because these measures appear by themselves in a single column.
3. Adding regional percentage of the whole measures (e.g., % of Total Revenue) makes it easy to assess the relative performance of regions.
4. Adding per-product percentage of the whole revenue measures makes it easy to assess relative product performance.
5. Adding derived measures—profit, average revenue per salesperson, and average order size—provides useful ways to assess performance that the vice president would not be able to calculate on the fly in her head.

Appendix H   ANSWERS TO PRACTICE IN
GRAPH DESIGN

Exercise #1

Here's the original graph:

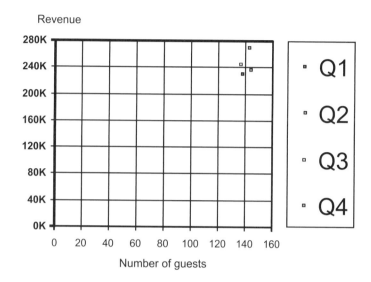

List each of the problems that you detect in the design of this graph:

1. The fundamental problem with this graph is that it is the wrong type. Scatter plots display correlations, but four data points aren't enough to establish whether or not a correlation exists. At best, this is a time-series relationship that involves two sets of quantitative values expressed in two different units of measure (revenue in dollars and number of guests as a count).
2. The four data points are too small to be easily distinguishable.
3. The grid lines are not necessary, and, even if they were, they are too visually prominent.
4. The axes are too visually prominent, standing out more than the data objects.
5. There is no reason for the labels along the vertical scale to be visually prominent.
6. There is no reason for the legend containing the quarters to be visually prominent. Details that make it prominent include enclosure in a border and a much bigger font than anything else.
7. The title of the vertical axis (Revenue) is not aligned with the labels.
8. The title of the graph (Cedar Isle) uses a different font from the rest for no apparent reason.
9. There is no date or other indicator of the year from which the information was taken nor is there any indicator of the currency.

Now, suggest a solution to each of these problems:

1. Use three line graphs, arranged vertically, with quarters along the horizontal axis. On the vertical axes, one graph should have revenue dollars, one should have number of guests, and one should have average revenue per guest in dollars.
2. If data points were appropriate, they should be larger.
3. Eliminate the grid lines.
4. Visually mute the axis lines compared to the data.
5. Make the revenue dollar text labels no more visually prominent than the other labels.
6. Eliminate the legend.
7. Align the axis titles with the labels.
8. Use the same font throughout the graph.
9. Add a date and an indicator of the currency.

Here's an example of what the solution might look like when designed effectively:

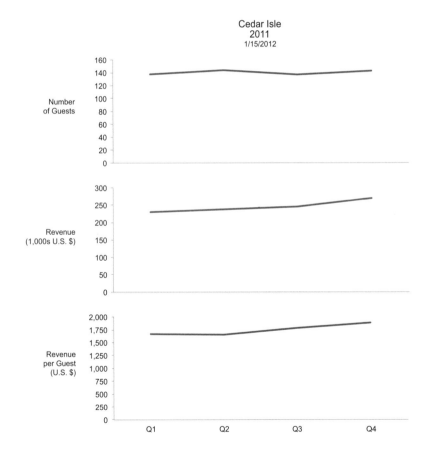

What this shows is that none of these values changed dramatically during the year, and that although revenues increased each quarter, the number of guests and revenue per guest did not increase. A zero-based scale was used in this series of graphs to maintain some consistency between them, but, because changes from quarter to quarter were so slight, the pattern of change does not stand out clearly. This can be solved by adding a second series of graphs that are scaled to begin just below the lowest value of each. Both sets together, as shown below, provide a more thorough view of the data:

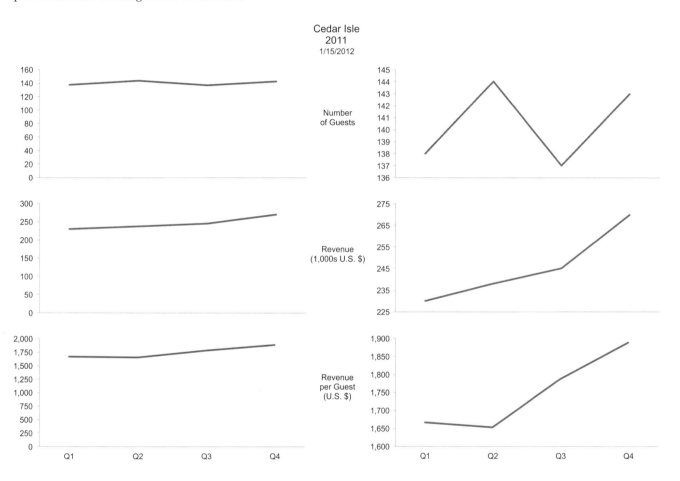

Exercise #2

Here's the original graph:

Kansas

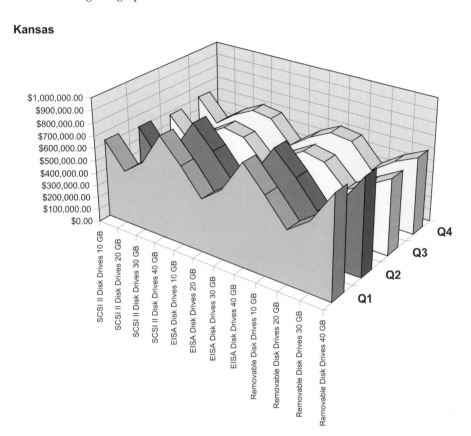

List each of the problems that you detect in the design of this graph:

1. The most significant problem with this graph is that lines have been used to connect values that are discrete and that have no intimate connection. It is meaningful to connect values along an interval scale, but not along a nominal or ordinal scale.
2. This graph is 3-D, which makes it impossible to read.
3. Encoding the values as 3-D objects also makes them more difficult to read.
4. The graph contains too many categorical items. It is difficult to display 12 products by four quarters in a single graph.
5. There are too many tick marks and labels along the vertical axis.
6. The numbers along the vertical axis should not include decimals. Including decimals makes the numbers less efficient to read and suggests a level of numeric precision that isn't available.
7. The labels for the quarters do not need to be more prominent than the other sets of labels.
8. The vertical orientation of the labels along the horizontal axis makes them difficult to read.
9. The fill pattern on the side and back walls adds no value but certainly adds distraction.
10. The grid lines add no value.

Now, suggest a solution to each of these problems:

1. Place time on the X axis so the four quarters will be connected by the lines rather than the products.
2. Restrict the graph to two dimensions.
3. Encode this time-series information as a separate line for each product.
4. Construct a series of graphs rather than trying to encode all 12 products in a single graph. Because the products group nicely into three sets of four products each (SCSI II, EISA, and Removable disk drives), each of these groups can be displayed in a single graph to make comparisons easier.
5. Reduce the number of tick marks and labels along the quantitative scale.
6. Remove the decimals from the numbers on the quantitative scale.
7. Make all the scale line labels equal in visual prominence.
8. Remove the products from the X axis, which eliminates the need to change their orientation.
9. Eliminate the third dimension, which removes the walls, eliminating the need to remove the distracting fill pattern.
10. Remove the grid lines.

Here's an example of what the graph might look like when designed effectively:

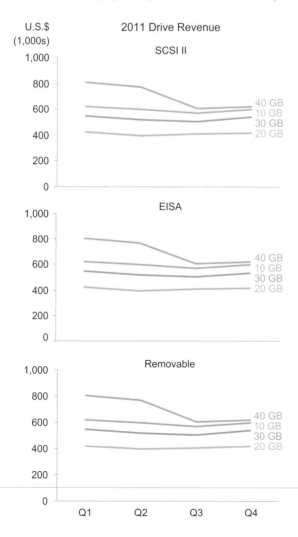

Exercise #3

Here's the original graph:

List each of the problems that you detect in the design of this graph:

1. The percentages in the 2Q of 2003 don't add up to 100%.

2. For each year, five pie charts are used to show what a single pie chart can display. Even if a single pie chart had been used for each year, however, the market shares in a given year would be difficult to compare.

3. Changes in part-to-whole relationships from one point in time to another are difficult to compare when displayed as pie charts.

4. The two points in time are arranged one above the other rather than in the natural side-by-side arrangement.

5. The black background behind gray text makes the values unnecessarily difficult to read.

6. Overall, by caring more about making this data display look like two network switches, one stacked on the other, the designer sacrificed clear communication.

7. Nothing in the graphic features the main point of the story, which is that increases in market share of the top four vendors came at the expense of smaller rivals.

Now, suggest a solution to each of these problems:

1. Correct the percentages in Q2 2003 to add up to 100%. (I did this by increasing the value of Others, which I renamed Smaller Rivals.)

2. If pie charts provided the best means of displaying this information, two pie charts would be sufficient, but we're going to find a better solution.

3. Replace the pie charts either with a line graph with one line per company, which connects two values—Q2 2003 on the left and Q2 2004 on the right—or with a bar graph that displays the changes in values from Q2 2003 to Q2 2004.

4. Arrange the two periods of time horizontally with Q2 2003 on the left and Q2 2004 on the right.

5. Either represent changes in market share as lines connecting Q2 2003 to Q2 2004 to make the patterns of change easier to see and compare, or display only the difference between the two points in time as bars.
6. If it is useful to include a picture of network switches to add visual interest to the graph, place it next to the graph rather than incorporating the graph into it.
7. Feature data related to the smaller rivals and perhaps even incorporate the fact that increases in the market share of the top vendors came at the expense of smaller rivals into the title or a note.

Here's an example of what the graph might look like when designed effectively as a line graph:

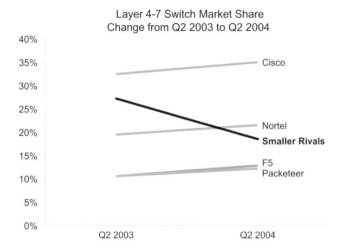

Here's an example of what the graph might look like when designed effectively as a bar graph:

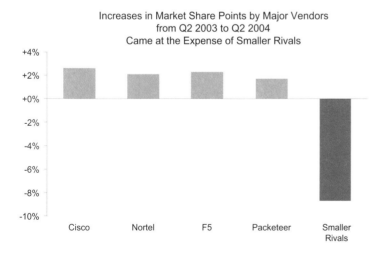

Exercise #4

Here's the original graph:

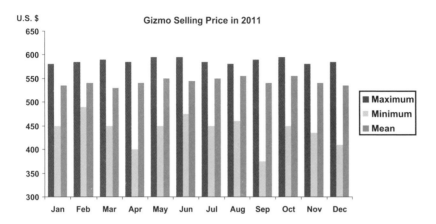

List each of the problems that you detect in the design of this graph:

1. Using separate bars to encode the monthly maximum, minimum, and mean values is not suitable for visualizing the trend of the mean through time or its relationship to the full range of monthly values.
2. The most important set of values is the means, yet the means are less visually prominent than the maximum values.
3. Sequencing the mean values after the minimum values is awkward. It is more natural to position the means between the maximum and minimum values.
4. Far too many components of this graph are visually prominent. It is hard to look at. The information does not stand out. The most visually prominent component of the entire graph is the legend because of its enclosure within borders.
5. The tick marks along the horizontal axis are not necessary.

Now, suggest a solution to each of these problems:

1. The trend of the mean values as they march through time should be encoded as a line. The maximum and minimum values may also be encoded as simple boxes.
2. Highlight the line that encodes the mean values.
3. Using boxes to encode the maximum and minimum values eliminates the problem associated with the awkward sequence of the bars.
4. Visually mute the supporting components.
5. Eliminate the tick marks along the categorical axis.

Here's an example of what the graph might look like when designed effectively:

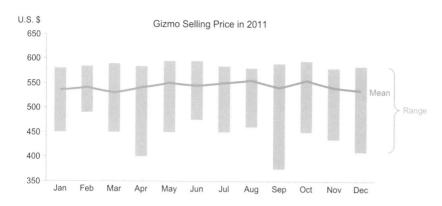

Exercise #5

Answers don't apply to this exercise.

Exercise #6

Here's one potential design solution for the proposed scenario:

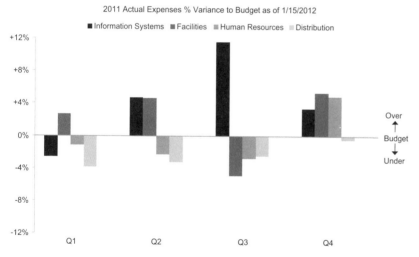

Note: Departments are sorted in order of overall variance from budget for the year, from worst to best.

Some of the highlights of this solution are:

1. Because the purpose of this graph is to display a deviation relationship, it is ideal to convert the actual dollar expenses to a percentage difference from the budget. Setting the baseline of the quantitative scale to the budget value of 0% makes deviations easy to see as bars extending up or down from that baseline.

2. The values could have been encoded as lines rather than bars, but the bars place more emphasis on the distinct quarterly values than on the trend over time.

The design in this example solution is not the only design that would work
effectively. Even if your design is significantly different, compare its merits to
those of the example solution to see how well it did in meeting the objectives
specified in the scenario.

Exercise #7

Here's one potential design solution for the proposed scenario:

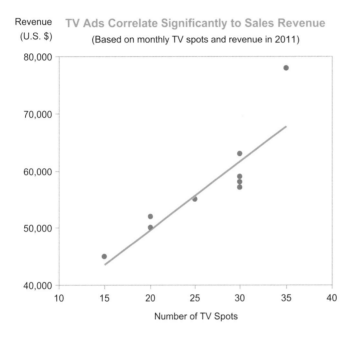

Some of the highlights of this solution are:

1. Because the purpose of this graph is to display a correlation between two
   variables, not to track the values across time, a scatter plot is an ideal
   choice.
2. The use of a trend line makes the strong positive correlation obvious.
3. The use of the contrasting green hue for the title and the trend line draws
   readers' eyes to the two most important pieces of information: the
   conclusion of the analysis and the information that supports that
   conclusion.
4. Subtle borders around the data region form a useful enclosure to keep the
   residual value of $78,000 in the upper right corner from seeming
   extraneous.

Exercise #8

Here's one potential design solution for the proposed scenario:

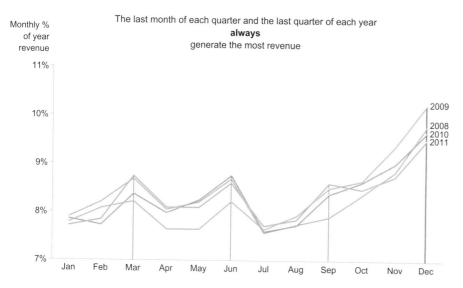

Some of the highlights of this solution are:

1. Because the message requires that the same months of each of the four years be compared to one another, the most effective layout involves assigning each of the years to the same set of 12 months rather than arranging the four years in sequence from left to right along the horizontal axis.

2. I've converted each month's revenue into a percentage of the entire year's revenue, which makes it easier to compare monthly contributions to the year by eliminating differences in revenues from year to year.

3. A clear title, which points out the important message contained in the data, makes the graph very easy to interpret.

4. Reference lines mark the last month of each quarter, especially in the final quarter, thereby focusing attention on these points in time to support the message.

5. Labeling the lines directly eliminates the need for a legend.

Appendix I   USEFUL COLOR PALETTES

Colors that were used in this book to break values into different categorical groups were almost always selected from the following three palettes. The light palette was used for large data-encoding objects, such as bars and boxes, the medium palette was used for small data-encoding objects, such as data points and lines, and the dark and bright palette was used to highlight particular items, such as a particular bar. To make it easy to reproduce these colors in your own software, the RGB values are provided for each color below.

Light

R	G	B
140	140	140
136	189	230
251	178	88
144	205	151
246	170	201
191	165	84
188	153	199
237	221	70
240	126	110

Medium

R	G	B
77	77	77
93	165	218
250	164	58
96	189	104
241	124	176
178	145	47
178	118	178
222	207	63
241	88	84

Dark & Bright

R	G	B
0	0	0
38	93	171
223	92	36
5	151	72
229	18	111
157	114	42
123	58	150
199	180	46
203	32	39

INDEX

About the Author . . .

STEPHEN FEW has been working for nearly 30 years as a teacher, consultant, and innovator in the fields of business intelligence and information design. Since 2003, when he founded the consultancy Perceptual Edge, he has focused exclusively on data visualization. He teaches, speaks, and consults internationally with organizations of all types. More than anyone else working in data visualization today, Stephen is respected and sought the world over for his ability to make data visualization accessible in simple and practical ways to anyone who wants to effectively understand and communicate the important stories that reside in quantitative information.

In addition to this book, Stephen has written two other popular books: *Information Dashboard Design: The Effective Visual Communication of Data* (2006) and *Now You See It: Simple Visualization Techniques for Quantitative Analysis* (2009). He also writes the quarterly *Visual Business Intelligence Newsletter* and teaches his *Visual Business Intelligence Workshop* publicly in several countries each year.

When he isn't working, he can usually be found—in or around his home in Berkeley, California—lost in a good book, savoring a fine wine, hiking in the hills, or instigating an animated discussion about the meaning of life with good friends.

For current information about Stephen and his work, go to www.PerceptualEdge.com.